Australian Endangered Species: Eucalypts

L.D. Pryor
assisted by
J.D. Briggs

Special Publication [5]

Australian National Parks and Wildlife Service

The views expressed in this publication are not necessarily those of the Director, National Parks and Wildlife, The Minister for Home Affairs and Environment, or the Commonwealth Government.

ISBN 0-642-89679-8

Cover photographs:

Front: *Eucalyptus benthamii* (Camden White Gum) — growing along the banks of the Nepean River near Camden. This species is threatened by agricultural clearing up to the river edge, invasion of exotic plant species and the inhibition of regeneration by stock grazing.

Photo: N. Hermes

Back: *Eucalyptus rhodantha* (Rose Gum)
Photo: L. D. Pryor

FOREWORD

Only a small proportion of Australia is forest covered. Nevertheless trees of the genus *Eucalyptus* are a dominant feature of the landscape in most of the more densely settled regions of Australia. In Australia eucalypts range from the cold alpine regions to the tropical north with its extremes of wet and dry, and along watercourses in the arid inland. The two tallest hardwood trees in the world are eucalypts, the Mountain Ash and the Karri.

Because of their abundance, variety and widespread distribution, eucalypts play a critical role in the life cycle of many native animals and plants. Forests of eucalypts protect water catchments and provide most of the native timber used in commerce.

Eucalypts have been introduced into many countries for amenity and wood production purposes. Frequently they exceed native tree species in growth and timber production.

Australians have a special international responsibility to protect representative natural stands of eucalypt species, some are more vulnerable than may appear. Of approximately 550 known species of eucalypts, 124, or nearly a quarter, are at some risk.

In 1977 the Australian National Parks and Wildlife Service engaged a leading world authority on eucalypts, Emeritus Professor L.D. Pryor, to examine the current position of endangered and rare eucalypts throughout Australia. This publication is a result of that study.

It is hoped that this publication, along with "The Natural Distribution of *Eucalyptus* in Australia" by G.M. Chippendale, also published by the Australian National Parks and Wildlife Service, will contribute to a better understanding of eucalypts and help to ensure the survival of all species.

Professor J.D. Ovington,
Director,
Australian National Parks &
Wildlife Service,
P.O. Box 636,
Canberra City A.C.T. 2601

TABLES AND FIGURES

AUSTRALIAN ENDANGERED SPECIES: EUCALYPTS

By

L.D. PRYOR

Assisted by **J.D. Briggs**

CONTENTS

INTRODUCTION

Trees of eucalypts are often the most prominent plants in Australian forest and woodland environments. Because eucalypts form such a dominant element of the flora the fate of many associated plant and animal species depends on that of eucalypts.

The merit of preserving biological species from extinction, has been stated frequently. In the Australian scene this has been set out succinctly by J.D. Ovington's 'Australian Endangered Species; Mammals, Birds and Reptiles' (1978).

The history of biological evolution records the elimination of many species and this occurred without the involvement of man. However it seems that man is now the prime cause of the elimination of species which is occurring at a much faster rate than previously. The growth of human population in the last few centuries has brought an explosive phase in the biological history of the earth. The phenomenal increase over the last three or four hundred years in the capacity of people to mould landscapes means the rate of species annihilation will probably reach an all-time peak. There are many good reasons to seek to cushion this change and to reduce it so far as possible, but this cannot be done without cost and some sacrifice.

Australia has been called by Len Webb and his colleagues 'The Last of Lands' (Webb et al. 1969) Biologically this is true because only comparatively recently has this country been exposed to the huge biological changes experienced in the rest of the world. These changes have developed rapidly in the last few centuries in Australia. When Europeans settled in Australia, a mere 200 years ago, the only major human impact, that of the Aboriginals, had attained relative stability over a period of occupation of at least 40,000 years. The advent of Europeans was biologically cataclysmic. So far as the survival of individual species is concerned this impact has been largely adverse. However for some, such as brush-tailed possums, certain kangaroo species, some beetles and birds and a number of plants, it has meant an increase in numbers. A great number of plant species have been much reduced in number, but in spite of this few have become extinct. The total numbers of individuals in populations are not in themselves a major factor in species survival, although there are of course critical minimal levels for all species.

It appears that no single eucalypt species has yet been lost due to the agency of Europeans. However, several species are in a highly precarious position and many are vulnerable in the light of the present day pace of change. We can predict that other species in fifty or hundred years time may be in the same precarious position although secure enough for the present. Attention needs to be directed to those species in a very vulnerable state or clearly at risk in the next few decades. The status of the remaining eucalypt species should be monitored to indicate where and when species protection is necessary. No doubt for some species there will not be any threat in the foreseeable future, but for perhaps half of the total array of eucalypt species, developments in the next century may reduce them to critical population levels. The risk is not only from the direct impact of modern society but also from indirect causes such as the inhibition of regeneration which is often much more insidious in effect and much less readily appreciated by the casual observer.

A program of preservation has a cost, sometimes large. However the very pressures which generate the threat are often associated with greater affluence in society suggesting that it may be possible to meet the cost if society so chooses.

THE EUCALYPT RESOURCE

Eucalypts are some of Australia's most notable and distinctive plants. This large genus of trees is almost entirely confined to Australia, with a few species in Papua and one in New Guinea, Sulawesi and Mindanao. There are two species in Timor and the adjoining Indonesian Islands (Figure 1). In general eucalypts are the most important trees in the Australian landscape and have important ecological and land use roles. Eucalypts are often planted in the warmer regions of the world, where they are much valued in the local economy. This is particularly so in developing countries, since in cultivation good growth is maintained even under rather difficult environmental conditions.

Altogether, there are some 550 species and subspecific taxa with probably more yet to be discovered. Some species are widespread in Australia, whereas others are very highly restricted in their natural occurrence. The majority are trees, some of which are very large indeed, while a few are shrubs. Because this biological resource is almost totally confined to Australia, the preservation of the various species in their natural habitat has both national and international implications. Apart from its economic and scientific values the genus has strong aesthetic appeal.

Characteristics of Species Distribution

In Australia the distribution of a species is usually regarded as that area over which the species occurred naturally at the time of European settlement. Earlier there may have been quite different distributions; however these distributions are a matter for speculation.

The common practice in describing the distribution of any species is to delineate the area by connecting the outer limits of all known occurrences. This then indicates a unit area through which the species is distributed. However because species occur in discrete stands making a mosaic with other species, the notional area of distribution is almost always an area of discontinuous patches of greater or lesser extent of the species. On this basis a species having a wide distribution with many stands would be called a common, widespread species. Alternatively a species may have a similarly wide distribution but with few stands and would then be an infrequent, restricted, or even rare species but with a wide distribution. If the geographic extent of the species is not wide it might be described as of limited distribution with either a continuous or disjunct distribution within these limits.

Species of limited geographic distribution and with few stands are called rare and are vulnerable to elimination because of their limited extent alone, apart from any other factors.

There are no precise terms used to describe the distribution of eucalypt species but the use of the words widespread, restricted, disjunct, common, rare are used to convey their original status and are adequate for most purposes.

Rarity of Species

Much interest in maintaining species will be directed to those regarded as 'rare'. A species may be considered rare for factors other than that the total population is quite limited.

Some species are considered rare where little is known about their distribution. Where detailed surveys have been undertaken the location of almost all occurrences may be known but it is usually impossible to be sure that the last stand has been found. It is always prudent to assume they may be more extant than any survey estimate allows. A very high percentage of recorded occurrences are at sites reached at least by a four-wheel drive vehicle whilst very few can be reached only on foot or horse. The implication is that the discovery of stands of species has depended very much on the ease of access. Therefore there are likely to be more occurrences of a species than have been recorded simply because of the difficulty and time involved in visiting some of the less accessible areas where they may occur.

Another category of 'rare' species is one which has only recently been recognised and described as a distinct species. Until a specific survey is made, the extent of these cannot be assessed with precision (see footnote Table 1). Particularly in Western Australia, it is likely that species recently described and presently known only from the type locality or perhaps one or two other places, are likely to be found to be more widespread when a survey is made based on knowledge of the taxon.

Features of Species of Restricted Distribution

While the ranges of many species of eucalypts extend over hundreds or even a thousand kilometres, a considerable number are of quite restricted distribution. There are some eighty species of very limited distribution. Current land use has already further limited some of these. Where such species occupy only ten to a few hundred hectares, the risk of total loss may be acute. For some species, protective measures are now urgently needed if they are to survive. Where they are on private land used for pastoral or other agricultural pursuits, there is a likelihood of removal of mature trees as well as restriction of regeneration. Often some trees survive along road verges which are generally under the control of a local government authority. Up to now local government managers generally have been unaware of the value and restricted nature of these species so that suitable protection programmes have not been implemented. Sometimes, during road upgrading, alignments can be varied to protect small surviving stands of restricted species. Whilst amendments to plans will involve greater costs, this is part of the price to be paid for conservation programmes of this type. Most people would probably agree that the amounts involved are fully justified in terms of the conservation objectives, although at times it may not be a charge which should be borne fully by the local authority.

Apart from directly threatened species, there are others of very limited occurrence in more remote and undeveloped areas. These species could easily be destroyed should the land use be changed e.g. with the introduction of mining and such engineering works as communications facilities, highways, powerlines or reservoirs. These species could be exterminated even in a national park. Such species can be described as potentially threatened since the whole population could be quickly eliminated. These and the foregoing species are listed in Table 1.

Other species of limited distribution are faced with the same risks as previously described but the threat to them is not acute because some natural stands still remain. If the rate of change of the last half century is maintained for the next 50 years a good proportion of the species in this group could be put at some risk. Land use change in relation to such species should be monitored and appropriate action initiated as necessary. These species of limited distribution are listed in Table 5.

The allocation of species to the different categories in Table 1 and 5 is essentialy arbitrary but the subdivisions provide a satisfactory basis for planning a preservation programme.

CAUSES OF REDUCTION OF EUCALYPT POPULATIONS

In the 200 years of European settlement in Australia the pattern of natural vegetation has greatly changed. The principal agent of change has been clearing of land for agriculture. This has led to a dramatic reduction of the *Eucalyptus* populations in many areas. This contraction is characteristic of farming and grazing areas of southern Australia and especially in the original vegetation types generally known as woodland or mallee. Dry or wet eucalypt sclerophyll forests have been relatively less changed since the new land use pattern has not always involved a complete removal of the original vegetation. However these ecosystems have also often been substantially modified. In general, current land use does not present the same acute threat to the eucalypt species of sclerophyll forests in the foreseeable future, even though there has been considerable publicity and criticism of the practice of converting such forest to other uses including forest plantations.

Of the 42 million hectares of productive or potentially productive eucalypt forest used for timber harvesting, only about 0.7 million hectares of plantation has been established. Plantation forestry has not always involved the direct conversion of eucalypt sclerophyll forest, since often there has been an intermediate phase in which the land was used for agriculture or grazing with associated tree clearing. Moreover, large areas of these sclerophyll forest types are in state forests managed for eucalypt timber production where species survival is seldom in question while the current sustained yield practice is maintained.

Where there has been a great reduction of eucalypt populations due to agriculture and grazing but the species still persist widely, albeit in much reduced numbers, a more subtle risk is present. This is where the grazing regime of domestic animals (and sometimes vermin) inhibits effective eucalypt regeneration. Any assessment of vulnerability of species must take the regeneration capacity of the species into account. Nevertheless the eucalypt species of the woodlands of southern Australia cannot yet be considered at risk so far as species survival is concerned. Monitoring of change by long term extensive repetitive survey is required, rather than any urgent action to ensure species survival. In comparison, species of highly restricted distribution are often directly threatened with extinction for a number of similar and other reasons. Being confined to limited areas they are unable to endure these pressures which pose no immediate threat to widespread species.

No described eucalypt species, as far as is known, has been lost totally from areas of natural occurrence. A few are in an extremely precarious position but even *Eucalyptus steedmanii*, which had not been seen on a natural site for many years, was rediscovered recently. In fact several stands of this species were found and it is not immediately at risk unless there is a change in land use. If appropriate action is taken now there is a good chance that all known eucalypt species can be preserved for the future *in situ* in a natural habitat.

For species confined to soils and situations where land use has eliminated most of the original populations and continues to erode the remainder, there is great threat of extinction. Few eucalypt species are in this category. The main ones are *E. morrisbyi*, *E. froggattii*, *E. yarraensis*, *E. parvifolia* and *E. rhodantha*. In a similar but slightly less precarious position, because there are larger surviving populations, or because new land use is not so threatening, are *E. argophloia, E. cneorifolia, E. conglomerata, E. crenulata, E. kartzoffiana, E. kitsoniana, E. lane-poolei, E. michaeliana, E. neglecta, E. remota and E. stoatei.* A few other species of restricted distribution are threatened by urban development. Particular examples of this are *E. camfieldii* and *E. risdonii* and to some extent *E. curtisii.*

Another group of species having very restricted distributions is still found over most if not all of their original range. This is because they occupy relatively 'safe' sites, so far not committed to land uses which are destructive of the natural vegetation. Often they are already in parks, reserves or other land under some kind of public ownership. There is a potential threat from future community needs which may override even the most rigorous restraints on development. Such restricted species in state forests, while for the most part equally secure could become vulnerable if a management policy change were implemented. This could include converting the natural vegetation to another forest use or varying the composition of the forest to eliminate species currently of low economic value.

Because species of limited extent are potentially endangered, knowledge of this fact should help in the examination of future development proposals. Many species may not come under threat during the next century or even longer, but there is a lack of certainty in predicting their future and some will probably become at risk. The status of species in Table 4 should be reviewed sooner than a similar review for the widespread species. Perhaps this should be done during the next decade since, in addition to those factors listed above as being potentially damaging, more insidious factors are operating such as lack of regeneration and decline of mature trees through age, disease or pests. For example *E. nova-anglica* has suffered great decline in many areas due to 'dieback' in the last decade and populations of *E. raveretiana* are expected to be reduced substantially soon by dam construction.

THREATENED EUCALYPT ECOSYSTEMS

A distinction should be drawn between providing for the survival of eucalypt species and for the preservation of good samples of complete ecosystems which include eucalypts. For the latter, considerably larger areas are required and generally more intensive management is necessary. Currently the greatest lack of preserved ecosystems in eucalypt communities is in the woodlands of the southern two-thirds of Australia. Most of these systems have been converted to agricultural use and very few remain in a more or less natural state. An exception is the *E. albens* woodland of the lower Snowy River Valley, a portion of which is in the south-east corner of the Kosciusko National Park and the remainder in a reserve in adjoining Victoria. It would be a costly operation to set aside and reconstitute samples of such systems through much of southern Australia. Nevertheless, such an initiative could be at least as rewarding as some of the current forest conservation campaigns concerned with 'rainforest' and 'woodchips'.

MEANS FOR ENSURING SURVIVAL OF THREATENED SPECIES

The species critically threatened in the immediate future and perhaps until about the end of the century, are primarily those of very restricted distribution. Survival implies not only preventing the destruction of some existing stands but ensuring that the conditions to which they are subjected will permit natural regeneration as older trees die. The locations of stands of these species are on sites with various sorts of land tenure.

Where the site is in public ownership and designated as national park, flora reserve or one of the several categories aimed at nature preservation, or alternatively on state forest where management policy is aimed at sustained yield production from indigenous forest species, there is usually little immediate risk of loss of the stand. It is necessary of course that the management authorities concerned in each case should be aware of the existence of each species to enable them to take this into account in framing management plans.

The position is different where the stands are on land either privately owned or under the control of local government or other public authorities without a specific role in nature conservation. On privately owned land the owner may be quite unaware of the endangered status of the species. Those species most acutely threatened are the dozen or so which so far as is known occur only on privately owned land, perhaps with associated small roadside verges. Land of this type is managed with varying degrees of intensity for agriculture and grazing. Regeneration is generally minimal and fluctuates according to seasonal changes and to the intensity of agricultural management. A good deal of this change will be geared to the prevailing economic levels of rural activity, so that in a slack period of rural production survival of the species is likely to be favoured because the stimulus to land development will be low and also the intensity of grazing may be reduced.

Preservation requires continued surveillance. Those areas under private ownership are less likely to be effective for long term preservation because the attitude of individual owners, even if favourable now, may change or there may be changes in ownership. Without guidance, owners as a rule, will be unaware of the best means to ensure the survival of the stand. This management will almost always impose some cost on the owner, possibly well beyond the owners resources.

The steps necessary for the survival of rare species are to ensure that clearing of established trees does not take place and that the factors inhibiting regeneration are controlled. Different approaches to meet this need have been adopted by different States. Purchase of the appropriate portion of land where the cadastral subdivisional arrangements are suitable is one method that has been followed and management then becomes vested in a public authority. Alternatively some kind of financial subsidy might be provided to an owner in return for the setting up of a management system which achieves the desired end. This usually means designating a preservation area and fencing or otherwise withdrawing the area from general rural use.

In the case of local government authorities where the stands of rare eucalypts are situated on road verges or reserved areas such as travelling stock reserves, the attitude of the authority involved will vary. As with private ownership an underlying difficulty will always be the cost of achieving arrangements for preservation. Frequently the simplest way of road improvement is by widening the road with minor realignment. This often results in a major reduction of the surviving verge vegetation. In some cases a significant portion of the surviving stands of a rare species is to be found now on such verges, especially where there has been extensive clearing and development of adjoining areas. Engineering solutions to such problems are generally fairly simple but the legal and financial implications may be relatively complex. Given that the cost can be met, road engineers can usually design without difficulty an alternative road alignment to leave sections of the verge along the older way completely untouched. However this usually involves resumption of land privately owned possibly with additional costs such as added fencing or added direct costs in engineering construction. It is a matter for judgment as to whether the expense involved in specific cases is merited by the conservation objective and whether this is a fair charge against the local authority. Often it seems scarcely a fair charge to be met by the local authority totally since the preservation objective has a much wider implication in the flow of its benefits. At the same time some immediate benefits in the local scene may not necessarily be lacking.

UNUSUAL PROVENANCES

For widespread species there are often geographic outliers, mostly as small stands, which have particular biological interest. They are likely to have an unusual genetic makeup and may provide evidence of past history of the vegetation. In the eucalypts there are some notable examples of such stands, and some geographic areas have a higher proportion of these than others. For example in South Australia there are several species occurring in restricted stands there but which are widespread in the eastern States. The South Australian occurrences are separated from the major stands by some hundreds of kilometres. One of the most outstanding of these is an area *E. macrorhyncha* near Seven Hills in the Clare district of South Australia which has been incorporated in a substantial reserve. The nearest occurrence of this species is in the Grampians in Victoria. There is a similar situation for *E. rubida* in the Adelaide Hills. An isolated occurrence of *E. maculata* near Nowa Nowa in Victoria, and the occurrence of *E. diversicolor* at Karridale in Western Australia are further examples of this pattern. There are many levels of disjunction in occurrences of this kind (See Chippendale (1980) for further details on disjunctions) and a survey of these would be complex and require considerable time. The preservation of particular provenances is perhaps of somewhat less importance than the preservation of species themselves, but should rank very highly in future conservation considerations.

BIOLOGICAL CONSIDERATIONS FOR SPECIES WITH RESTRICTED DISTRIBUTION

A consideration of the species of highly restricted distribution leads to some speculation about their history. Eliminating from the list those species which are relatively closely related to other species, leaves about 50 highly distinctive species. They occur on various sites: some are rocky with skeletal soils; others are wet or poorly drained areas; some are on relatively highly fertile soils; others are on average well drained sites. A number of species in this category are associated with remnants of dissected lateritic topography. In systematic affinity they are distributed through the genus in the main taxonomic groups except for the genus Corymbia (See Table 2). This group has an essentially northern distribution with fewer highly restricted species than the size of the group would predict.

Vigour of growth in the field, and vigour as shown when raised under conditions of cultivation are very variable. The stature of trees of these species in their natural occurrence tends to be small rather than large, although a few are very large trees. The distinctness of the taxa and their distribution in such limited natural stands have the hallmarks of relicts. This view is heightened by the fact that a number have markedly disjunct distributions and that outlying provenances occur for some well known species.

There seem two possible causes for this feature of eucalypt distribution. Firstly the advent of man to Australia, over 40,000 years and perhaps 100,000 years ago, was accompanied by a change of fire characteristics in the habitat and so helped to bring about the present patterns of vegetation. Alternatively, a long term change in climate, known to have occurred from the evidence of past glaciation and other sources, would perhaps be an equal, or even more potent, factor in bringing about the present vegetation distributions. Possibly both explanations contributed to the modern distribution patterns.

In general the relict nature of the restricted species implies superior competitive ability by the 'successful' species. The role of interspecific competition within the genus is well illustrated by a comparison of distribution patterns observed in East Timor with those in a comparable ecological range in Australia. In East Timor there are only two *Eucalyptus* species, one is *E. urophylla* endemic to the lesser Sunda Islands and Timor and the other is *E. alba*. These occupy a range of sites from sea level to more than 3000m altitude. *E. alba* occurs from the coast to about 1500m and overlaps for about 1100m with *E. urophylla*. In the overlap *E. alba* occupies the poorer soils and harsher aspects. Above 1100m *E. urophylla* occupies all sites and aspects to 3000m with very considerable temperature and rainfall variations. It thus covers an environmental range which in Australia would provide habitats occupied by perhaps twenty or more different eucalypts. Imagine a hypothetical situation in Australia where the number of species is very limited as in Timor. *E. pauciflora* (s. lat.) for example occurs from the treeline to almost sea level between Bega and near the top of Mt Kosciusko. Presumably if it was one of the hypothetical two species in the absence of competition from other *Eucalyptus* species it would probably occupy most of the available microclimatic sites, determined according to altitude, aspect and general substrate conditions.

It appears that in the main eucalypt area of Australia, species are confined to their various sites, usually of relatively limited extent, primarily because of competition at the boundaries. This competition is with adjoining species which are able to out-perform their neighbours on sites which suit them better. This would explain why species like *E. benthamii* and *E. raveretiana* occur naturally on limited sites but are vigorous growers and display vigour in cultivation too. These species are found naturally on, and restricted to, levee banks and alluvial soils associated with river systems which are areas of relatively higher nutrient status than is usual. Presumably they require the better conditions to compete successfully against species which thrive on the adjoining poorer soils.

Looked at broadly on this basis, the total *Eucalyptus* population in Australia consists of those species having a wide range which might be described as the ones successful under present day conditions and those which are restricted to very limited habitats by the competitive effect of these successful species. Given a different climatic regime, one can imagine the balance being changed with some species at present relatively unsuccessful and strictly confined gaining a competitive advantage. They would then spread into some areas presently occupied by other species.

There is evidence of wetter and cooler conditions at times in the past, with oscillations to conditions even more arid than at present. Such large climatic variations, would have had a considerable effect on species distribution especially in southern Australia.

The relatively smaller topographic relief of northern Australia and proximity to the equator makes a similar effect less likely. There are relatively fewer restricted species in the north. However it must be remembered also that northern Australia has a smaller total eucalypt species population for comparable ecological situations, and there has been less species botanical investigation.

Species of restricted distribution are particularly characteristic of south-eastern Australia and south-west Western Australia. There is a preponderance of these species in the south-east where the topography is more diverse and where there are more sites which could become survival refuges for relict species.

If the species of very restricted distribution are properly regarded as relict species, the fact that they occur often on sites marginal or unsuitable for growth of adjoining 'successful' species has a further implication. These relict species have been reduced in area presumably by their inability to compete with the more successful species during changing environmental circumstances. During this process they may have been exposed to sites not previously available to them to which they were then able to adapt. In short they displayed a genetic plasticity sufficient to outstrip other species in at least the very small or difficult niches which they now occupy. This seems to fit the observed pattern with species such as *E. crenulata, E. sturgissiana, E. neglecta* and *E. parvifolia* on rather wet or cold sites. In rocky crevices or on skeletal soils it fits *E. scoparia, E pulverulenta, E. olsenii, E. pumila* and *E. mitchelliana* in the east and *E. insularis, E. caesia* and *E. acies* in the west.

On the other hand some relict species are simply tied to topographic remnants of old surfaces such as *E. squamosa* in the east, *E. remota* on Kangaroo Island and perhaps *E. lane-poolei* in the west. These species may have achieved their relict status simply because sites to which they were adapted were slowly eroded away and they were unable to adapt to the new surrounding conditions.

The very restricted distributions of another group of species includes *E. yarraensis, E. froggattii, E. benthamii, E. kartzoffiana,* and *E. argophloia.* These are characteristic of sites with better soils in each of their respective distribution areas and a similar reduction in the extent of such sites may have caused a contraction of the species populations.

The speculative search for factors causing changes in suitable habitats for particular species is generated partly by the predicted theoretical need to ensure evolutionary development for larger populations in the past than the present very small patches. Larger populations would be expected to permit the evolutionary emergence of these highly distinctive species. This distinctiveness is a feature of many presumed relict species and is displayed in the most telling way by species such as *E. olsenii, E. pumila, E. neglecta, E. crenulata, E. insularis* and *E. kruseana.*

ACTION TO SECURE PRESERVATION OF ENDANGERED EUCALYPT SPECIES

In virtually all cases, more detailed surveys must be the first step taken to ensure the preservation of threatened species. In some cases this has already commenced through the enthusiasm of interested individuals or organisations. Such surveys are likely to reveal additional stands and add to the knowledge of the extent of stands of these restricted species.

An important aspect for each species is to establish precisely the tenure of the land on which the stands are found. With this information it will be possible to assess the administrative alternatives to ensure the preservation of the species and possibly to secure the financial requirements to achieve the aim. In most cases participation by government authorities will be necessary.

TABLE 1 *Eucalyptus* species of highly restricted occurrence and endangered or potentially endangered

1. *E. acies*
2. *E. alpina*
3. *E. approximans* ssp. *approximans*
4. *E. aquilina*
5. *E. archeri*
6. *E. argophloia*
7. *E. badjensis*
8. *E. baeuerlenii*
9. *E. bakeri*
10. *E. barberi*
11. *E. beardiana***
12. *E. benthamii*
13. *E. brockwayi*
14. *E. burdettiana*
15. *E. burgessiana*
16. *E. caesia*
17. *E. calcicola*
18. *E. camfieldii*
19. *E. cneorifolia*
20. *E. conglomerata*
21. *E. cordata*
22. *E. coronata*
23. *E. crenulata*
24. *E. cupularis**
25. *E. curtisii*
26. *E. dendromorpha*
27. *E. desmondensis*
28. *E. ficifolia*
29. *E. fitzgeraldii**
30. *E. froggattii*
31. *E. georgei***
32. *E. gregsoniana*
33. *E. halophila***
34. *E. imlayensis***
35. *E. insularis*
36. *E. johnsoniana***
37. *E. kartzoffiana*
38. *E. kitsoniana*
39. *E. kruseana*
40. *E. lane-poolei*
41. *E. lansdowneana* ssp. *lansdowneana**
42. *E. largeana*
43. *E. luehmanniana*
44. *E. megacornuta*
45. *E. michaeliana*
46. *E. mitchelliana*
47. *E. morrisbyi*
48. *E. neglecta*
49. *E. nigra**
50. *E. olsenii*
51. *E. paliformis*
52. *E. parvifolia*
53. *E. pendens*
54. *E. pulverulenta*
55. *E. pumila*
56. *E. remota*
57. *E. rhodantha*
58. *E. risdonii*
59. *E. rummeryi*
60. *E. saxatilis*
61. *E. scoparia*
62. *E. sepulcralis*
63. *E. squamosa*
64. *E. steedmanii*
65. *E. stenostoma*
66. *E. stoatei*
67. *E. stowardii**
68. *E. sturgissiana*
69. *E. tetrapleura*
70. *E. triflora*
71. *E. woodwardii*
72. *E. xanthonema**
73. *E. yarraensis*

* Distribution not adequately known and in *E. nigra* taxonomic boundaries are not clear. These species are not discussed in the text.

** Species newly described and the distribution not adequately known. These species are not discussed in the text.

TABLE 2: *Eucalyptus* species listed on TABLE 1 as being endangered placed in systematic groups

Subgenus—CORYMBIA
Section—N/A

E. ficifolia
E. fitzgeraldii

Subgenus—EUDESMIA
Section—N/A

E. curtisii

Subgenus—MONOCALYPTUS
Section—N/A

E. acies
E. alpina
E. approximans ssp. *approximans*
E. aquilina
E. burgessiana
E. camfieldii
E. conglomerata
E. coronata

TABLE 2 (Cont'd)

	E. dendromorpha
	E. gregsoniana
	E. insularis
	E. johnsoniana
	E. luehmanniana
	E. mitchelliana
	E. nigra
	E. olsenii
	E. paliformis
	E. pendens
	E. remota
	E. risdonnii
	E. sepulcralis
	E. stenostoma
	E. triflora
Subgenus—SYMPHYOMYRTUS Section—BISECTARIA	*E. bakeri*
	E. beardiana
	E. brockwayi
	E. burdettiana
	E. caesia
	E. calicicola
	E. cneorifolia
	E. desmondensis
	E. halophila
	E. kruseana
	E. lane-poolei
	E. megacornuta
	E. rhodantha
	E. squamosa
	E. steedmanii
	E. stowardii
	E. xanthonema
Subgenus—SYMPHYOMYRTUS Section—DUMARIA	*E. georgei*
	E. stoatei
	E. woodwardii
Subgenus—SYMPHYOMYRTUS Section—EXSERTARIA	*E. cupularis*
	E. michaeliana
	E. pumila
Subgenus—SYMPHYOMYRTUS Section—MAIDENARIA	*E. archeri*
	E. baeuerlenii
	E. badjensis
	E. barberi
	E. benthamii
	E. cordata
	E. crenulata
	E. imlayensis
	E. kartzoffiana
	E. kitsoniana
	E. morrisbyi
	E. neglecta
	E. parvifolia
	E. pulverulenta
	E. saxatilis
	E. scoparia
	E. sturgissiana
	E. yarraensis

TABLE 2 (Cont'd)
Subgenus—SYMPHYOMYRTUS
Section—ADNATARIA

E. argophloia
E. froggattii
E. lansdowneana ssp. *lansdowneana*
E. largeana
E. rummeryi
E. tetrapleura

TABLE 3: *Eucalyptus* species listed on TABLE 1 as being endangered arranged by geographic location

QLD	SA	NSW	VIC	TAS	WA
E. approximans	*E. cneorifolia*	*E. approximans*	*E. alpina*	*E. archeri*	*E. acies*
ssp. *approximans*	*E. lansdowneana*	ssp. *approximans*	*E. crenulata*	*E. barberi*	*E. aquilina*
E. argophloia	ssp. *lansdowneana*	*E. badjensis*	*E. froggattii*	*E. cordata*	*E. beardiana*
E. bakeri	*E. remota*	*E. baeuerlenii*	*E. kitsoniana*	*E. morrisbyi*	*E. brockwayi*
E. conglomerata		*E. bakeri*	*E. mitchelliana*	*E. risdonii*	*E. burdettiana*
E. curtisii		*E. benthamii*	*E. neglecta*		*E. caesia*
E. scoparia		*E. burgessiana*	*E. saxatilis*		*E. calcicola*
		E. camfieldii	*E. yarraensis*		*E. coronata*
		E. dendromorpha			*E. cupularis*
		E. gregsoniana			*E. desmondens*
		E. imlayensis			*E. ficifolia*
		E. kartzoffiana			*E. fitzgeraldii*
		E. largeana			*E. georgei*
		E. leuhmanniana			*E. halophila*
		E. michaeliana			*E. insularis*
		E. nigra			*E. johnsoniana*
		E. olsenii			*E. kruseana*
		E. paliformis			*E. lane-poolei*
		E. parvifolia			*E. megacornuta*
		E. pulverulenta			*E. pendens*
		E. pumila			*E. rhodantha*
		E. rummeryi			*E. sepulcralis*
		E. squamosa			*E. steedmanii*
		E. stenostoma			*E. stoatei*
		E. sturgissiana			*E. stowardii*
		E. tetrapleura			*E. xanthonema*
		E. triflora			*E. woodwardii*
					NT
					—

TABLE 4: Assessment of degree of risk to species in TABLE 1 according to apparent land tenure of main areas on which existing stands occur.

A. Mainly on private land usually with some occurrence on associated roadside verges.
B. Mainly on public areas or unoccupied crown land but not primarily managed for preservation of vegetation.
C. Mainly on public lands such as parks, nature reserves or state forest where preservation of vegetation is likely to be a main aim of management.
D. Restricted species whose distribution is inadequately known at present to assess level of threat.

A	B	C	D
E. argophloia	*E. baeuerlenii*	*E. acies*	*E. beardiana*
E. benthamii	*E. bakeri*	*E. alpina*	*E. cupularis*
E. cneorifolia	*E. burgessiana*	*E. approximans*	*E. fitzgeraldii*
E. conglomerata	*E. desmondensis*	ssp. *approximans*	*E. georgei*
E. cordata	*E. kruseana*	*E. aquilina*	*E. gregsoniana*
E. crenulata	*E. lansdowneana*	*E. archeri*	*E. halophila*

TABLE 4 (Cont'd)

A	B	C	D
E. curtisii	ssp. *lansdowneana*	*E. badjensis*	*E. imlayensis*
E. froggattii	*E. luehmanniana*	*E. barberi*	*E. johnsoniana*
E. kartzoffiana	*E. megacornuta*	*E. brockwayi*	*E. nigra*
E. kitsoniana	*E. olsenii*	*E. burdettiana*	*E. stowardii*
E. lane-poolei	*E. paliformis*	*E. caesia*	*E. xanthonema*
E. michaeliana	*E. pendens*	*E. calcicola*	
E. morrisbyi	*E. risdonii*	*E. camfieldii*	
E. neglecta	*E. steedmanii*	*E. coronata*	
E. parvifolia	*E. stoatei*	*E. dendromorpha*	
E. pulverulenta	*E. sturgissiana*	*E. ficifolia*	
E. rhodantha	*E. tetrapleura*	*E. insularis*	
E. yarraensis	*E. triflora*	*E. largeana*	
	E. woodwardii	*E. mitchelliana*	
		E. pumila	
		E. remota	
		E. rummeryi	
		E. saxatilis	
		E. sepulcralis	
		E. scoparia	
		E. squamosa	
		E. stenostoma	

TABLE 5: *Eucalyptus* species of relatively limited extent which may be threatened in the near future. (These species are not detailed in the text).

1. *E. acaciiformis*
2. *E. albida*
3. *E. apiculata*
4. *E. beyeri*
5. *E. brachyandra*
6. *E. brevistylis*
7. *E. chapmaniana*
8. *E. chloroclada*
9. *E. cladocalyx*
10. *E. collina*
11. *E. comitae-vallis*
12. *E. cosmophylla*
13. *E. crucis*
14. *E. dawsonii*
15. *E. diptera*
16. *E. dongarraensis*
17. *E. dunnii*
18. *E. erythrocorys*
19. *E. fasciculosa*
20. *E. fruticetorum*
21. *E. glaucescens*
22. *E. goniantha*
23. *E. grossa*
24. *E. guilfoylei*
25. *E. gunnii*
26. *E. haematoxylon*
27. *E. jacksonii*
28. *E. johnstonii*
29. *E. kybeanensis*
30. *E. lehmannii*
31. *E. ligustrina*
32. *E. macarthurii*
33. *E. mckieana*
34. *E. melanoxylon*
35. *E. mooreana*
36. *E. moorei*
37. *E. multicaulis*
38. *E. nicholii*
39. *E. nitens*
40. *E. nova-anglica*
41. *E. pachycalyx*
42. *E. pachyloma*
43. *E. perriniana*
44. *E. pimpiniana*
45. *E. pyrocarpa*
46. *E. raveretiana*
47. *E. rigidula*
48. *E. rudderi*
49. *E. rupicola*
50. *E. sargentii*
51. *E. torelliana*
52. *E. umbrawarrensis*

TABLE 6: *Eucalyptus* species listed on TABLE 5 as being at risk in the future arranged by geograph location.

QLD	SA	NSW	VIC	TAS	WA
E. pachycalyx	*E. cladocalyx*	*E. acaciiformis*	*E. chapmaniana*	*E. gunnii*	*E. albida*
E. raveretiana	*E. cosmophylla*	*E. apiculata*	*E. fruticetorum*	*E. johnstonii*	*E. brachyandr*
E. torelliana	*E. fasciculosa*	*E. beyeri*	*E. glaucescens*	*E. perriniana*	*E. brevistylis*
	E. pimpiniana	*E. chapmaniana*	*E. kybeanensis*		*E. collina*
		E. chloroclada	*E. nitens*		*E. comitae-val*
		E. dawsonii	*E. perriniana*		*E. crucis*
		E. dunnii			*E. diptera*
		E. fruticetorum			*E. dongarraen*
		E. glaucescens			*E. erythrocory*
		E. kybeanensis			*E. goniantha*
		E. ligustrina			*E. grossa*
		E. macarthurii			*E. guilfoylei*
		E. mckieana			*E. haematoxyl*
		E. moorei			*E. jacksonii*
		E. multicaulis			*E. lehmannii*
		E. nicholii			*E. melanoxylo*
		E. nitens			*E. mooreana*
		E. nova-anglica			*E. pachyloma*
		E. perriniana			*E. rigidula*
		E. pyrocarpa			*E. sargentii*
		E. rudderi			**NT**
		E. rupicola			*E. brachyandr*
					E. umbrawarre

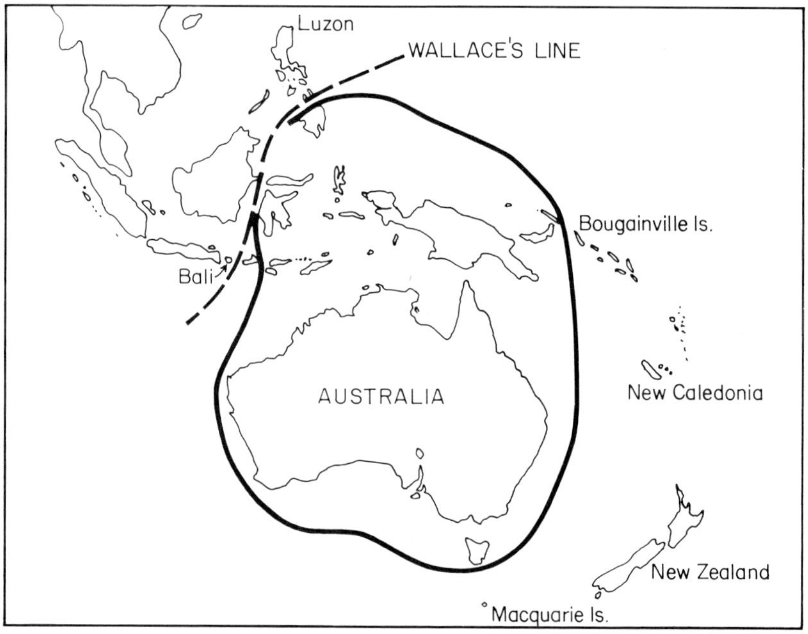

FIGURE 1

The limits of natural distribution of *Eucalyptus* in the Australasian region (solid line) showing its close conformity on the north-west boundary with Wallace's line which it does not cross to the Asian side. (After Pryor 1976).

FIGURE 2

The areas (stippled) in Australia in which *Eucalyptus* is prominent in the vegetation. It is not entirely absent from unshaded areas, especially in the arid central and central western regions, but when it does occur it is confined to limited sites such as seasonal water courses or rock clefts. These occurrences cannot be represented at the scale of the diagram. (Based on field map by J. A. Carnahan after Pryor 1976).

ACKNOWLEDGEMENTS

This work has been possible only because of most effective support by the Australian National Parks and Wildlife Service throughout its compilation.

Information and other assistance was freely given by all State Herbaria and Forestry authorities.

In particular, thanks are due to Dr J. W. Green, Curator of the Western Australian Herbarium, Dr L. A. S. Johnson, Director of the Royal Botanic Gardens and Herbarium, Sydney and Mr M. I. H. Brooker of the Division of Forest Research, C.S.I.R.O., Canberra. Mr G. M. Chippendale of the same Division provided not only information on distribution but also the basic maps.

Information on particular species was provided by many people, too numerous to name, which was very valuable in the course of the survey. We record our warm thanks to all of them.

The photographs were supplied by the Division of Forest Research, C.S.I.R.O., except where otherwise acknowledged.

L. D. P. & J. D. B.

REFERENCES AND FURTHER READING

Blakely, W. F. *A Key to the Eucalypts.* 3rd Ed. Commonwealth of Australia, Forestry and Timber Bureau, Canberra, 1965.

Boland, D. J., Brooker, M. I. H. and Turnbull, J. W. *Eucalyptus seed.* CSIRO, Canberra, 1980.

Hillis, W. E. and Brown, A. G. *Eucalypts for Wood Production.* CSIRO, Canberra, 1978.

Kelly, S. *Eucalypts Volume I.* Nelson, Melbourne, 1969.

Kelly, S. *Eucalypts Volume II.* Nelson, Melbourne, 1978.

Maiden, J. H. *A critical revision of the Genus* Eucalyptus Vols 1-8. Govt. Printer, Sydney, 1903-1931.

Ovington, J. D. *Australian Endangered Species: Mammals, Birds and Reptiles.* Cassells, Melbourne 1979.

Penfold, A. R. and Willis, J. L. *The Eucalypts.* Leonard Hill, London and Interscience Publishers, New York. 1961.

Pryor, L. D. and Johnson, L. A. S. *A Classification of the Eucalypts.* A.N.U., Canberra, 1971.

Pryor, L. D. *Biology of Eucalypts: Studies in Biology No. 61.* Edward Arnold, London, 1976.

Webb, L. J., Whitelock, D. and Brereton, J. Le Gay. *The Last of Lands.* Jacaranda Press, Milton, 1969.

Eucalyptus acies Brooker. **Woolbernup Mallee** (MABEJ F.T.S. 194)

This is a small mallee which is no more than about 2.5m tall. It is located on Woolbernup Hill in the Fitzgerald River National Park and on a couple of other ranges in the same general area.

Distribution

The species was first recognised on Woolbernup Hill and described by M.I.H. Brooker in 1970. The few other sites on which the species occurs are all clustered on the south coast of Western Australia between Bremer Bay and Hopetoun in about latitude 34°S. They occur at altitudes of about 300-450 m. Probably all of the original stands are still intact.

Description

It is a very small smooth-barked mallee occurring in dense clumps. A distinctive feature is the quadrangular shape of the small stems in cross-section and the stiff sharp edges. The scientific name of the species refers to this feature. It is one of a small group of species of the Monocalyptus "subgenus" from Western Australia and is somewhat similar to *E. preissiana*. It differs from this species in having buds and fruits in clusters of 7 and being considerably smaller.

Ecology

This species occupies hilltop sites on outcropping proterozoic quartzites which are very well drained and produce an acid soil It is found in the part of Western Australia which has a mild climate. The mean maximum temperature of the warmest month probably not above 24°C and the mean minimum temperature of the coolest month about 7°C. There are no frosts. It is in a zone where the rainfall is reliable with a marked winter peak and a dry summer. The mean annual rainfall is about 600mm. It occurs in small dense clumps and is usually associated with sclerophyll shrubs characteristic of that part of Western Australia.

Conservation

The species has so far been practically undisturbed except possibly from the indirect effects of fire which may not be very different from the pre-European situation. It occurs in the Fitzgerald River National Park and this is likely to ensure its preservation indefinitely. The management prescriptions in the park should take the conservation needs of the species into account. No other programme appears necessary to ensure the preservation of the species.

Special Features

E. acies is of particular interest because it is one of a small group of species endemic to Western Australia which are related to the eastern Australian Monocalyptus group. It is therefore of considerable biogeographic interest. It is also a distinct species of very limited occurrence, its prescence suggests that it is a relict species and may well have been more widespread when climatic conditions were wetter and milder than now.

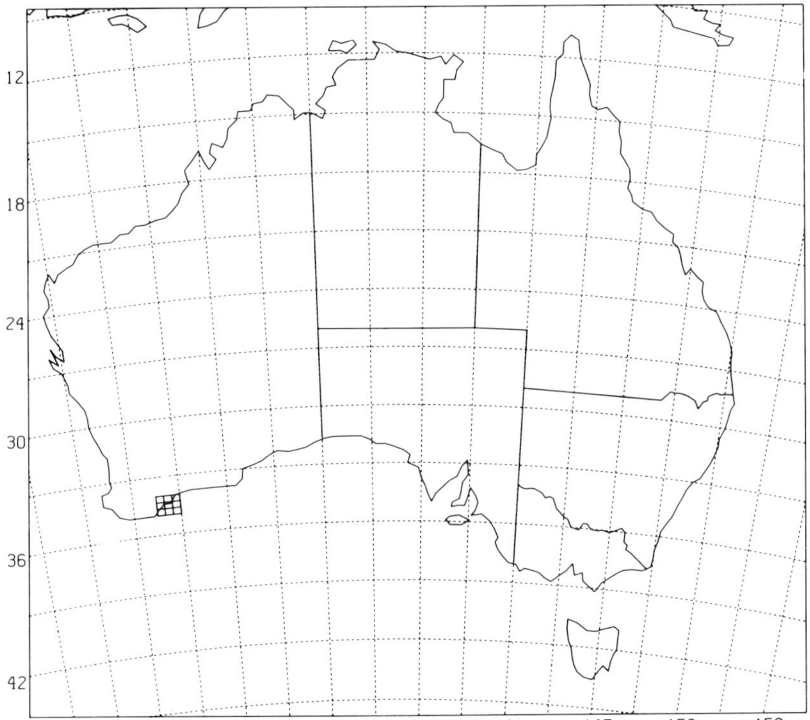
12
18
24
30
36
42
111
117
123
129
135
141
147
153
159

Eucalyptus alpina Lindl. **Grampians Gum** (MAHCC F.T.S. 79)

The name of the species is misleading since the Grampian Range of western Victoria could scarcely be described as an alpine location. It is a small bushy shrub, small tree or mallee. There seems to be some considerable diversity in the appearance of the tree in different stands which would merit further field and herbarium study.

Distribution

E. alpina has a quite restricted distribution in the Grampian range of western Victoria. It occurs on rocky sites on hilltops all of which are apparently located either in national parks or forest reserves. The largest population known occupies about 10 hectares at Mirranatwa Gap where it occurs along the roadside. A small part of this stand was lost during road construction some years ago.

Description

It is a small stringybark related to *E. baxteri* from which it differs in its buds, fruits and leaves. The juvenile leaves on the seedlings are characteristic of the stringybarks, being quite densely covered with "stellate" hairs. The adult leaves are broadly ovate lanceolate and quite thick with a dull dark green colour. The buds are tightly clustered and with a markedly verrucose appearance. Buds of three or four different ages can be seen developing on the plant at the one time. In the mallee forms of the species, the bark often decorticates completely, whereas on the small tree forms, a distinct stringybark persists on the main trunks.

Ecology

It occupies rocky sites at about 1200m elevation on peaks of the Grampian Ranges. It is found down to altitudes of about 400m at Mirranatwa Gap. These sites are all about latitude 37°S where there is a cool sub-humid climate with temperatures ranging from a mean summer maximum of about 23°C to a mean minimum for the coldest month below zero in winter. About 60 or 70 frosts per year are usual and the mean annual rainfall is about 800-1000mm with a winter peak.

Conservation

The species, because of its occurrence on rocky hilltop sites, has been subject to little disturbance and probably most of the original stands are still intact. Since they are located in national parks they are reasonably secure at present.

Special Features

The diversity in form of the species is striking. It merits more study to determine the nature of the variation within the species and also the species inter-relationship with *E. baxteri* and perhaps *E. obliqua*. It has been planted as a small ornamental shrub. Its unusual form, which makes it quite distinct from other stringybark species is a notable feature of this species. Also its isolated location on the outlying Grampian Ranges of western Victoria separated by considerable distances from other elevated land in South Australia and central Victoria gives it a considerable biogeographic interest.

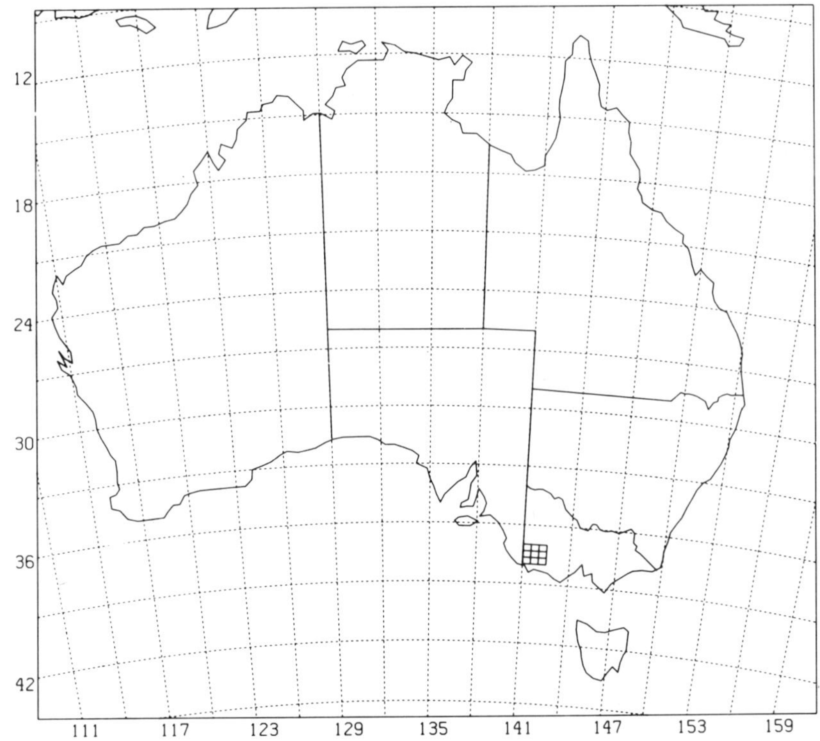
12
18
24
30
36
42
111
117
123
129
135
141
147
153
159

Eucalyptus approximans Maiden **Barren Mountain Mallee** (MAKIK F.T.S. 143)

The species is a rather small whip-stick mallee of from 2-5m in height. It is found in small patches on isolated rocky sites and never in a very substantial stand.

Distribution

The species occurs at Barren Mountain and near Mt. Norman and Castle Rock just on the NSW border in Queensland, and a few other similar locations in the same general region including Gibralter Range, NSW and Mt. Barney in Queensland.

Description

The species has two distinct sub-species which have not been formally described. The names *E. approximans* and *E. codonocarpa* have been given to them if they are to be regarded as distinct species. The difference between the two is mainly in the bell-shaped fruit of sub-species *E. codonocarpa* from which the name is derived. It is a smooth-barked whip-stick mallee with many stems arising from the common base. The leaves are rather narrow and long with a deep green colour which is rather distinctive. It has some affinity with *E. stricta* which is also a variable species with distinct sub-species found further south in NSW. The buds have quite a short operculum with a rounded top and are rather minutely verrucose.

Ecology

It occurs on rocky hilltops or slopes usually of massive granites often with large bare areas with the tree growing in fissures in and around the boulders. It occurs at altitudes around 1300m in areas near the edges of the northern tablelands of NSW and southern Queensland where misty rain is a frequent feature of the climate. The mean maximum temperature of the warmest month is about 24°C and the mean minimum of the coolest month about -2°C. There are numerous frosts during the year probably from 60 to 100 and the mean annual rainfall is about 1000mm with a summer maximum. There are occasionally light falls of snow. The occurrences at Mt. Norman are in a national park and at Barren Mountain very close to the boundary of the Meringal National Park in landscape which is unlikely to be disturbed. They are under little threat from any foreseen activity. Because of the way in which the species occurs and the general inaccessibility of the sites on which it is found it is quite likely that a more detailed survey would disclose additional stands on suitable sites.

Special Features

The variation in the species indicated by the recognition of two sub-species indicates the somewhat unusual feature of the plant. It is related to *E. stricta* and with several others which have a similar relationship such as *E. dendromorpha* and *E. obtusiflora*. The species is of considerable interest in the taxonomic sense because it presents a pattern of variation which is not easily explained by conventional taxonomic means. The status of the sub-species and other forms of it and related groups will probably not be fully disclosed until much more extensive study has been carried out. This will involve experiment in examining the developmental phases and seedling characteristics of different stands. The sites to which the species is confined suggests that under present climatic conditions it is reduced to the few areas which have a suitable climate and are free from the competitive stress provided by other more vigorous species of adjoining sites.

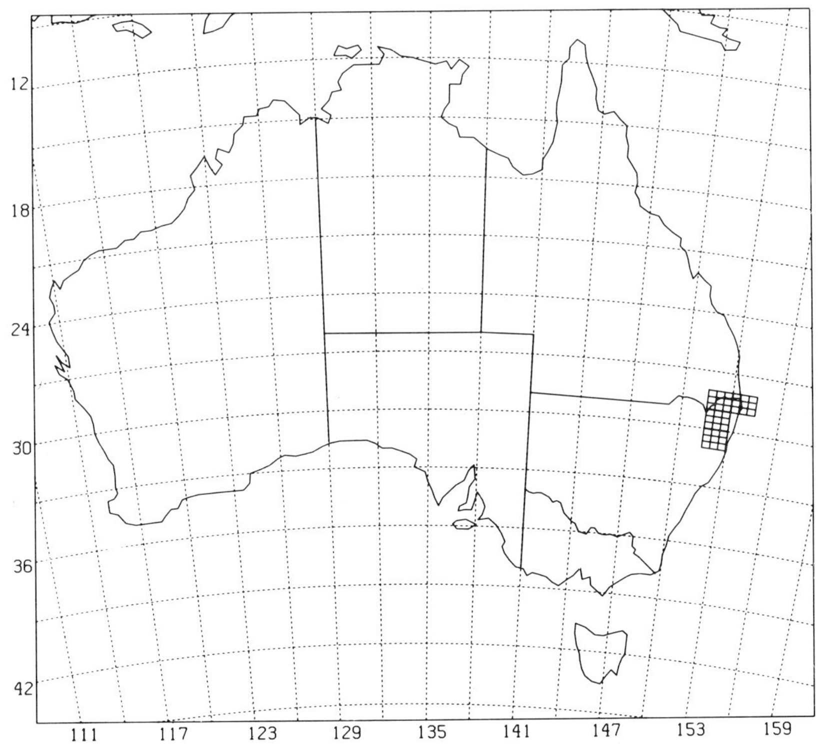

12
18
24
30
36
42
111
117
123
129
135
141
147
153
159

Eucalyptus aquilina Brooker **Mt Le Grande Mallee** (MABCC F.T.S. 181)

This eucalypt is a very small many-stemmed mallee which grows up to a height of about 4m.

Distribution

The species was recognised as distinct only recently when it was described by Brooker, although it had been collected as early as 1803 by Robert Brown during Flinders' voyage in that part of Western Australia. As far as is known it is confined to the Frenchman's Cap-Mt Le Grande area, Western Australia, which is within the Mt Le Grande National Park, in latitude 34°S and at an altitude of about 100m. It is likely that all the original stands are intact.

Description

E. aquilina is related to *E. coronata* and more remotely to *E. megacarpa*. In many ways it is similar to *E. coronata* and occurs geographically not far from the sites where that species is found. There is a major difference however in the operculum and a substantial difference in the fruits, although the buds and fruits of both species are quite large and in clusters of 3. It is a small many stemmed mallee forming dense thickets and is often associated in the field with spearwood mallee, *E. dorotoxylon* and usually with dense shrubs of many genera.

Ecology

The very limited occurrence of the plant does not allow identification of any special features of the site on which it occurs. The species is found on watercourses on moderate slopes which are rocky sites with skeletal soils derived from granite or gneiss. The climate is mild with a mean maximum temperature not above 25°C and the mean minimum of the coolest month not below 7°C. It is unlikely that there are any frosts in the area. The rainfall is reliable at around 600m annually with a marked winter peak.

Conservation

All known stands are within the Mt Le Grande National Park and the sites on which it occurs are not prominent. It seems unlikely that there is any significant risk of the loss of the species and it may be still other sites may be found if detailed survey is carried out. The species prefered habitat is quite limited along this coastal part of Western Australia.

Special Features

The taxonomic affinity of the species is of interest. In many ways it is very similar to *E. coronata* and the two species occur on similar sites and in a similar climatic zone. However *E. coronata* has a number of distinct distinguishing features. *E. aquilina* appears to be a relict species, the populations of which have been reduced almost to the point of extinction due perhaps to climatic changes in the past.

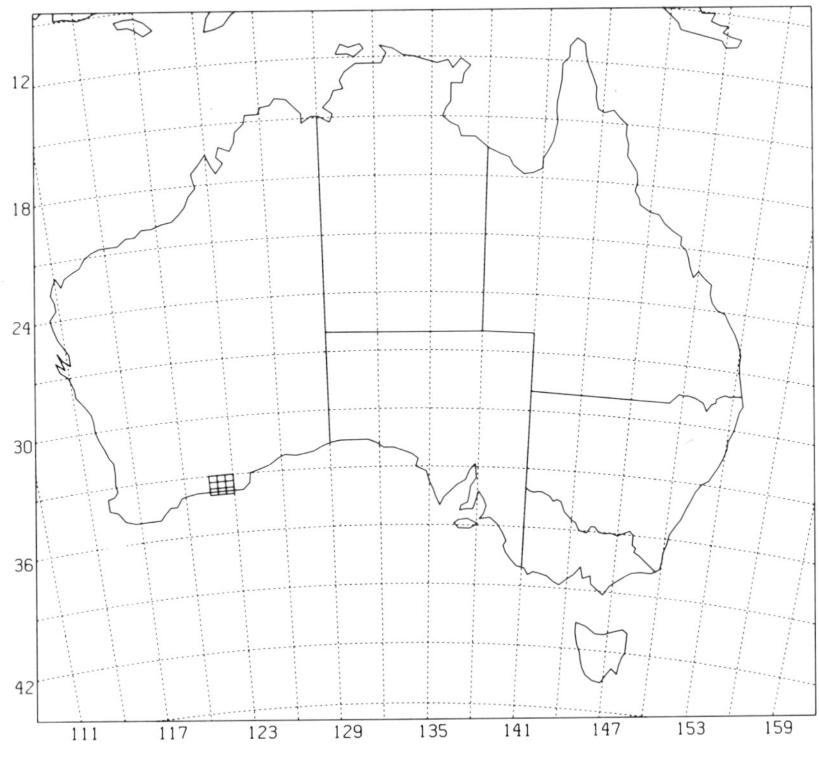
12
18
24
30
36
42
111
117
123
129
135
141
147
153
159

Eucalyptus archeri Maiden & Blakely **Archers Gum** (SPINIB F.T.S. 54)

E. archeri is regarded as *E. gunnii* sub-species *archeri* by Pryor & Johnston although it has not yet been formally reduced to the rank of sub-species. It is very closely related to *E. gunnii* which is not an endangered species.

E. archeri is a moderate sized tree on the tiers of north-western Tasmania.

Distribution

It is distributed in rather small patches on the edges of various tiers, especially in the Ben Lomond National Park. It is also found at a number of other locations in northwest Tasmania, for example near the Poatina head race tunnel.

Description

E. archeri is a small tree often reduced to little more than a shrub on hard rocky sites. It is a smooth barked gum quite similar to *E. gunnii* but with smaller buds and fruits which nevertheless are in clusters of 3. The juvenile leaves are the same orbicular, glaucous type as in *E. gunnii.*

Ecology

The tree occupies sites with shallow soil and rocky outcrops and it is often found mixed with *E. coccifera.* In a number of places it extends over substantial area with the usual associated small shrubs and herbaceous species. The mean maximum temperature of the warmest month is about 20° and the mean minimum of the coolest month about 0°C.

Conservation

It represents a variation of *E. gunnii* sufficiently distinct to merit preservation. It is located in the Ben Lomond National Park and also on sites in the Poatina area which are unlikely to be threatened significantly by any foreseen form of land use. Except for the prevention of destructive land use leading to the clearing of the species no particular programme is necessary to ensure its preservation. The distribution of the species should be noted and a more precise survey carried out to indicate its full extent. This species may be found to be considerably more extensive than at present recorded when the more inaccessible sites are examined.

Special Features

Its relationship with *E. gunnii* is interesting and this in turn with *E. morrisbyi.* One view is that *E. gunnii* in the broad sense, covered much more of Tasmania during periods of glaciation and that the present fragmented distribution of the two sub-species of *E. gunnii* and *E. morrisbyi* itself represents the final development following climatic fluctuations. While the growth habit of this sub-species may make it useful for some purposes, *E. gunnii* is more suitable for wood production or shelter in cold climates. It is not difficult to raise and cultivate *E. archeri.*

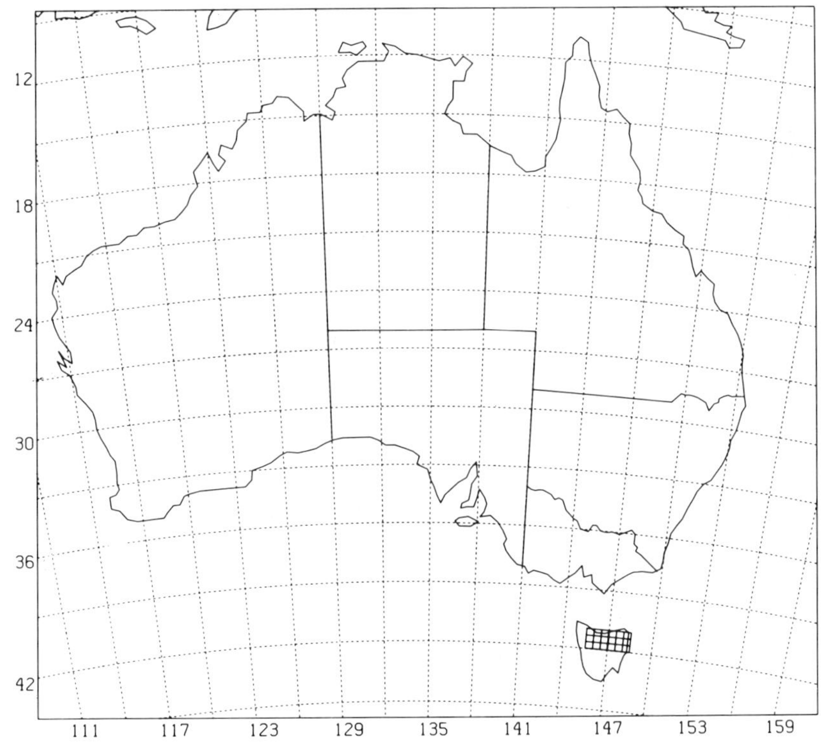
12
18
24
30
36
42
111
117
123
129
135
141
147
153
159

Eucalyptus argophloia Blakely **Queensland Western White Gum** (SUNAA F.T.S. 18)

The common name is appropriate for this silvery white barked tree. It is usually a splendid tall straight tree and is found in southern Queensland. Although called a gum it is one of the "box" group of eucalypts.

Distribution

It is in the Chinchilla-Jandowae area of southeastern Queensland and especially in the area known as Pelican. The total extent of the species was never large. It occupies soils favoured for agriculture and most of the original trees have been cleared. The surviving stands are few and scattered. They are now mainly regrowth following earlier clearing. Mature trees are mostly in patches of limited extent on privately occupied land or on roadside verges. There may be a small number in state forests but most are on alienated land.

Description

E. argophloia is a mostly smooth barked tree occasionally with a little fibrous "box" bark near the base of the trunk. It has quite narrow linear juvenile leaves, the usual terminal inflorescence and capsules with valves sunken below the rim of the fruit like *E. bosistoana*. Also like this species, to which it is quite closely related *E. argophloia* usually has 6 or 7 loculi in the ovary and a similar number of valves of the fruit. This is a distinctive feature.

Ecology

The tree occupies sites on lower slopes with soils of heavier texture than those at higher locations. The sites may be very moist or even temporarily waterlogged in wet seasons.

It is a long lived tree forming a tall woodland often in pure stands but at times with a discontinuous understorey of *Acacia harpophylla*, Brigalow. It regenerates well following clearing.

The species occurs in latitude 26°-26.25°S at about 300m altitude. there are usually 10-15 frosts a year and with mean maximum temperature of 34°C in the hottest month and the mean minimum of the coldest month about 5°C. Mean annual rainfall is about 650mm with a summer peak.

Conservation

Preservation of the species will require a three point management programme; to keep for as long as possible some mature trees, to maintain younger regrowth stands and also to secure further regeneration. A detailed survey of the area through which it is known to be or likely to have been distributed is an essential prerequisite in developing a precise plan.

Special Features

The species has potential value for amenity use as well as for production planting. It is both a distinctive and beautiful tree.

Its affinity with *E. bosistoana* which occurs south of Sydney in NSW and in eastern Victoria suggests a common origin. The great geographic separation of the natural stands of the two species by some 800km probably implies a long term separation of the original stock and subsequent divergence to the present two species.

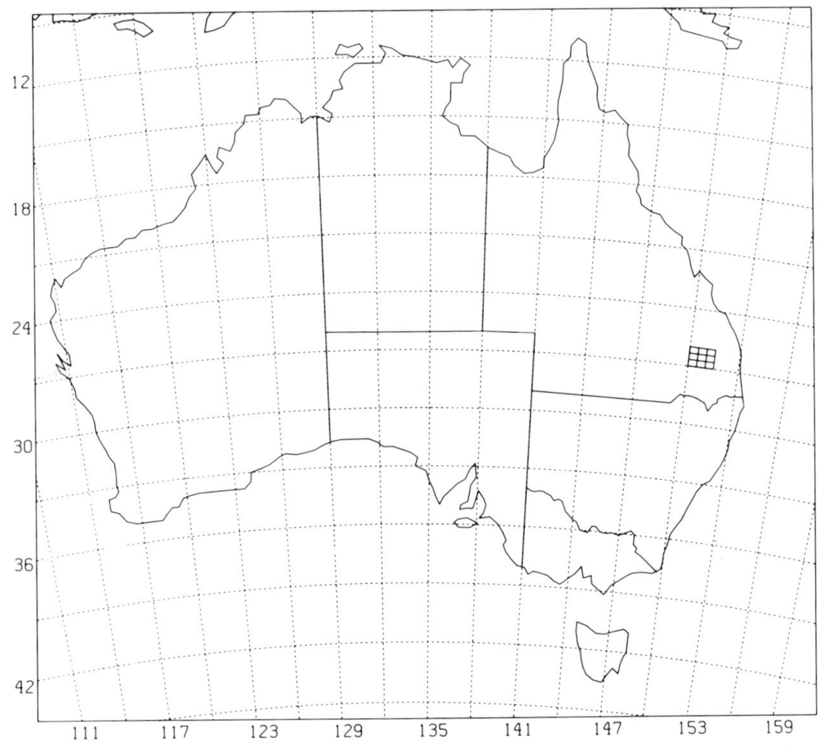
12
18
24
30
36
42
111
117
123
129
135
141
147
153
159

Eucalyptus badjensis de Beuz. et Welch **Big Badja Gum** (SPIKN F.T.S. 118)

This is a tall tree of moist habitats and good forest soil.

Distribution

It occurs in small stands separated from one another in a zone along the eastern edge of the southern tablelands of NSW from the vicinity of Big Badja Mountain south to Brown Mountain and beyond. Because of the habitat which this species prefers is limited the total extent is not large.

Description

This species is a tall tree up to 35m in height and with a stem diameter at breast height of 1.5m although on poorer sites it might be less than 25m in height. Some sub-fibrous rough bark persists with ribbons on the lower part of the trunk but the upper part is smooth and decorticating. It is quite closely related to *E. viminalis* having buds and fruits in clusters of 3 but, unlike this species the fruit has a sharply edged rim. The lateral buds in the clusters of 3 are placed rather obliquely rather than in the more cruciform arrangement common in *E. viminalis.*

Ecology

Big Badja Gum occurs in a sub-humid climatic zone of narrow extent which has mild summer temperatures and cool to cold winters. The mean maximum temperature of the warmest month is about 22°C whereas the mean minimum of the coldest month is around zero. There are probably numerous frosts perhaps up to 100 a year. The mean annual rainfall is about 1000mm and rather well distributed throughout the year with a substantial number of mountain mists. The best stands are on well developed fertile soils although it does extend onto harder sites from time to time. It is frequently mixed with *E. fastigata.* and *E. obliqua* and sometimes in the vicinity of *E. nitens.* On poorer sites it may be found near *E. elata, E. radiata, E. sieberi* and *E. smithii.* It is a component of typical wet sclerophyll forest with tall shrubs in the understorey vegetation often including *Acacia dealbata.*

Conservation

The full extent of *E. badjensis* is not well known because the recognition of it in the field has been rather delayed and precise surveys have not been carried out. It is likely to occur more extensively than the existing records indicate although it is still a species of very limited distribution. Much of the area in which it occurs is state forest and conservation would best be served by building into the prescription of forest management the preservation of the species in some suitably sized tracts. Until the distribution is more clearly known it is not certain whether it occurs in any national park or areas proposed as national parks. It is similar in many respects to other tall forest species and regenerates well. Providing the objects of management are to preserve the species in selected localities, this should present no particular problems as long as the natural forest is maintained.

Special Features

The fact that it is rather similar to *E. viminalis* has probably delayed its recognition although it was recognised when first described that the timber was of better quality than *E. viminalis.* Little is known about its performance in cultivation although it is not difficult to raise from seed. Under present management the species is not under much threat but if there were a change which involved replacing the eucalypt forest where it occurs with other forms of land use, for example the establishment of pine plantation, this could immediately change its status.

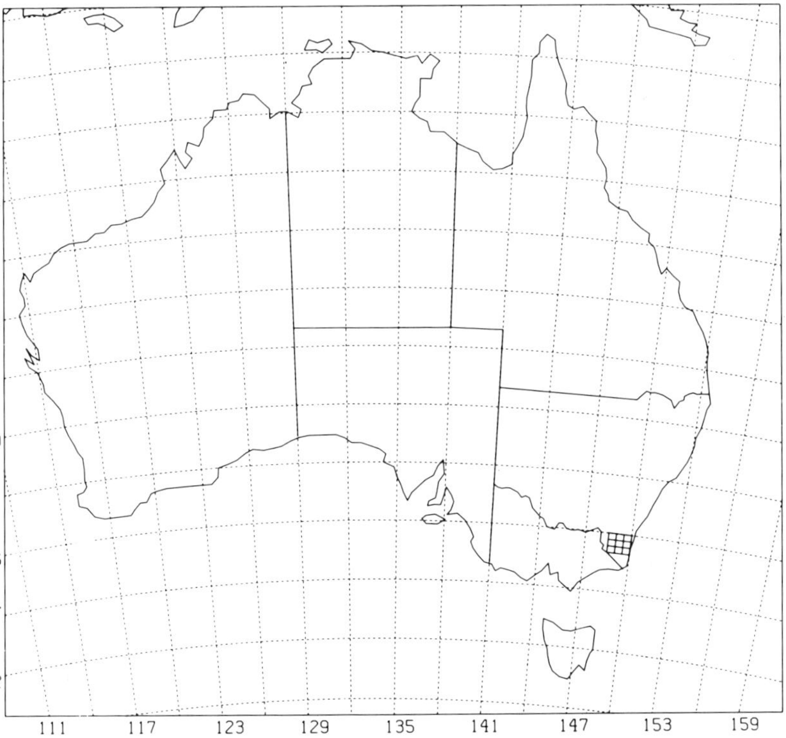
12
18
24
30
36
42
111
117
123
129
135
141
147
153
159

Eucalyptus baeuerlenii F. Muell. **Baeuerlen's Gum** (SPIKO F.T.S. 92)

This is a small tree only occurring in two localities west and south of Syndey in NSW.

Distribution

The tree occurs in two isolated localities—one on Sugarloaf Mountain in Monga State Forest at the top of the Clyde Mountain and the second at Wentworth Falls in the Wentworth Falls Nature Reserve. The two sites are separated by some 400km and there are no known stands between these two nor are there any other recorded stands elsewhere. it is therefore a species which has a slightly restricted distribution and with a substantial disjunction of the two stands.

Description

Baeuerlen's Gum is a slender tree up to about 20m in height with a smooth deciduous bark. At times it may be reduced to a whipstick mallee. The buds and fruits are in clusters of 3 but it differs from its nearest relatives such as *E. viminalis* and *E. badjensis* by having a conical or even rostrate operculum.

Ecology

The sites where the species occurs are rocky and it is in fissures and soils associated with these rocky outcrops. It occurs at some 1000m altitude and in the absence of a meteorological record one can only guess that the mean temperature of the warmest month is probably not above 27°C and the mean minimum temperature of the coolest probably about 5°C. Frosts occur on perhaps 50 days a year and the rainfall is from about 800-1000mm. The soils tend to be skeletal.

Conservation

The nature of the sites in which the species occurs and the location in a reserve and state forest implies that in spite of its restricted distribution it would only be threatened with extinction in circumstances which are not readily foreseen and which would be quite exceptional. Both of the sites from which it is recorded are fairly accessible by road although the Sugarloaf Mountain locality has to be reached for the last 800m by foot. This suggests that there may well be other sites of very restricted occurrence in rugged country between the two extremes which have so far not been found. Nevertheless while more is becoming known of the eucalypts, it has not yet been recorded from other localities, and it may be that there are no other occurrences to be discovered.

Special Features

This, along with a few other species, is of particular interest because of its disjunct distribution—that is to say, two limited occurrences separated by a substantial distance with no intermediate stands. There can be no doubt that they have had a common origin and are parts of the same species. The separation implies that they are vestigial remnants of what must have been at some earlier stage a wider distribution, although not necessarily continuous at any one time. Evidence from species with this kind of distribution is of great interest in speculations about the origin and evolutionary history of the Australian vegetation.

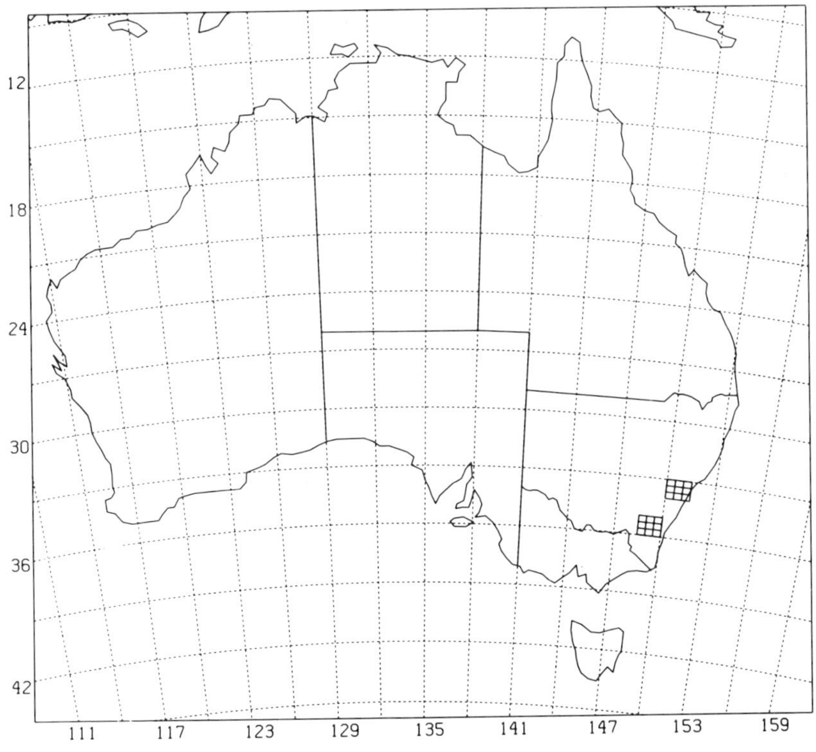
12
18
24
30
36
42
111
117
123
129
135
141
147
153
159

Eucalyptus bakeri Maiden **Bakers Mallee** (SIR:E F.T.S. 145)

Bakers Mallee is often a small tree up to 14m in height but it also occurs in the mallee form with many slender stems. It is found mainly in Queensland.

Distribution

This species is mainly found over a fairly wide area in southeastern Queensland some 150km from the coast, with a separate disjunct area in central Queensland, in which patches occur. The bigger of these occurrences is in an area some 500 by 400km and actually straddles the dividing range. Some stands occur in adjacent areas of NSW. In spite of the wide area of distribution it is confined to a mosaic of discrete patches within this area on sites suitable for mallee.

Description

This species is called a mallee but is often a single stemmed tree up to 14m in height and as much as 50cm in diameter. Elsewhere it occurs as a typical mallee of smaller dimension with many slender stems from the common underground lignotuber. Although it has rather dull foliage and in the field can be confused easily with *E. viridis* it is quite distinct from this species. The fruits are small with exserted valves and the buds have a rather long operculum. In the tree form a platey rough bark persists for some distance on the butt.

Ecology

The region in which it occurs is warm and sub-humid with the mean maximum temperature of the hottest month around 32°C and the mean minimum of the coolest month around 7°C. A small number of frosts occur each year and the mean annual rainfall is about 600mm with a pronounced summer peak. It occurs on slight rises. The soils on these rises are of poorer quality than the surrounding areas, often lateritic or even skeletal. Although Bakers mallee is found over a substantial area it is nowhere common and occurs only in small patches. It is often associated with *E. viridis* and *E. caleyi* and a number of *Acacias* and *Callitris.*

Conservation

Because of the nature of the soils on which it occurs Bakers Mallee is probably not as much under threat as some of the other species of restricted distribution. There are many occurrences at present and work unconnected with agricultural or pastoral practices, such as roads or engineering works, would not be likely to eliminate many stands of the species since these are spread over considerable distances. More precise survey is necessary to determine whether there are stands of adequate character and condition in reserved areas to meet preservation needs.

Special Features

The nearest relative of the species in *E. jutsonii* and its sub-species which are found in Western Australia. It belongs to a group of eucalypts which have a common feature bisected cotyledons in the newly germinated seedling. Most of the species in this group, which are quite numerous, are confined to Western Australia and only three reach the east coast of Australia. These are *E. pachycalyx* of north Queensland, *E. squamosa* of the Sydney district and Bakers Mallee. The development of many species of the group in Western Australia and what would appear to be three relict species in eastern Australia suggests that this species is part of a record of past changes and different climates across Australia.

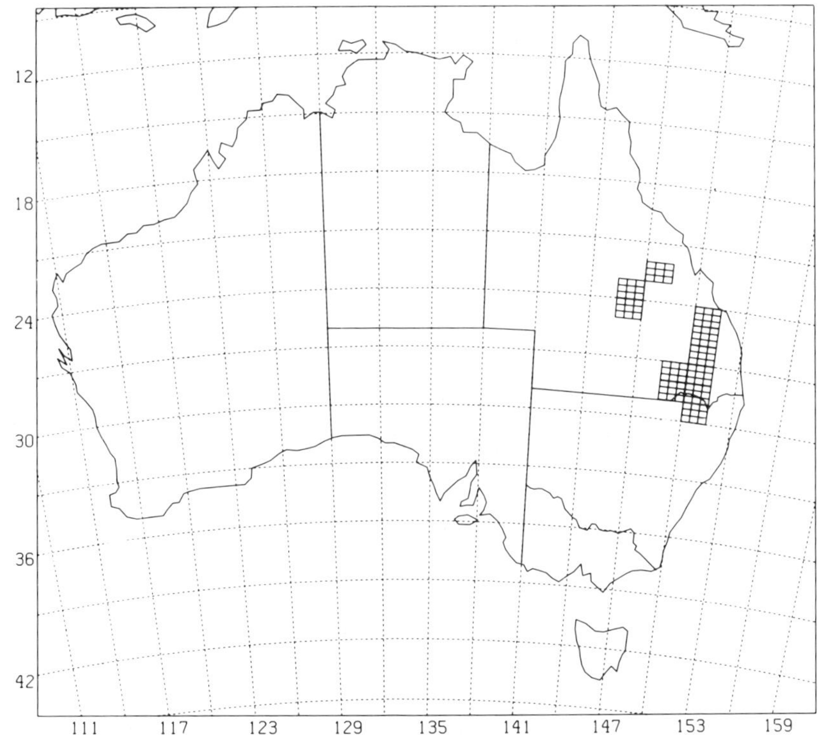
12
18
24
30
36
42
111
117
123
129
135
141
147
153
159

Eucalyptus barberi L. Johnson et D. Blaxell **Barber's Gum** (SPEAF F.T.S. 152)

This eucalypt is a small straggly crooked tree, sometimes with a mallee form, which occurs in the central east coast of Tasmania.

Distribution

Barber's Gum is very restricted in its distribution. It is found on low rocky hills in the Cranbrook area of Tasmania northeast of Hobart. It is in approximately latitude 42°S and at altitudes of from 100-200m. The sites on which it grows are quite limited in extent and consequently the species itself is of very limited distribution.

Description

It is a gum fairly closely related to *E. ovata* and *E. camphora* but nevertheless a quite distinct species. It has rather coarse foliage and the buds and fruits are in clusters of 7. The fruit is more cylindrical than either *E. ovata* or *E. camphora* and the juvenile leaves rather smaller than in those species.

Ecology

Barber's Gum grows on sites which are rocky ridges formed from granite or dolerite with reasonably well drained shallow black humus soils. The mean maximum temperature of the warmest month is about 22°C and the winters are cool with the mean minimum of the coldest month about 3°C. Because of closeness to the coast, frosts are not severe and probably average about 15 a year. The mean annual rainfall is about 650mm. There may be light snow falls of short duration at times in spring. It occurs mainly in small stands on its own but at times associated with White Peppermint, *E. pulchella* and Tasmanian Blue Gum, *E. globulus*. Swamp Gum, *E. ovata* is found on adjacent less well drained sites and an array of shrub species of typical Australian genera like *Melaleuca, Leptospermum* and *Hakea* are found associated with it.

Conservation

Because of the very limited nature of the sites and the small total amount of this species, its existence was overlooked until it was first recognised as distinct by the late Professor H.N. Barber. Most of the known occurrences are on privately owned property and some adjoining roadside verges. Because of the nature of the sites on which it occurs, it has been excluded from agricultural and pastoral development amd may not be affected much by these activities in the future. Adequate conservation however will only be assured if there is reservation of suitable stands. To select appropriate areas, a full survey of the distribution of the species is needed.

Special Features

Presumably because it is of limited extent and although it is a quite distinct species, Barber's Gum was overlooked in botanical collections for a long time. It may have been passed over as a variant of *E. ovata*. It is a striking example of the feature which prevails in *Eucalyptus* in which species of a quite distinct character are found only in very limited sites. Presumably they have developed with much wider distributions and are now confined to limited sites.

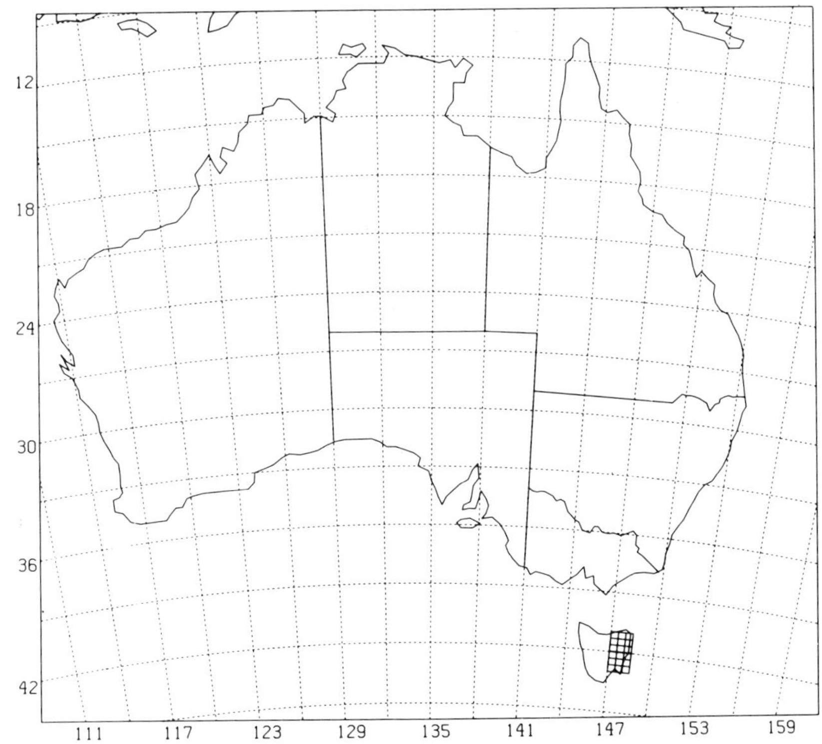
12
18
24
30
36
42
111
117
123
129
135
141
147
153
159

Eucalyptus benthamii Maiden et Cambage **Camden White Gum** (SPIKQ F.T.S. 57)

Camden White Gum is a splendid tree of large dimensions with a dense rather umbrageous crown.

Distribution

It is found in quite limited areas a little to the west of Sydney on river flats and levee banks of the Nepean River and its tributaries including those now within the Warragamba Dam. Originally it occurred over an area probably 100km long by 40km wide in a latitude of about 34°S and low altitudes of around 100m or less. The original stands were distributed between Camden and Penrith along the Nepean River and also along Cox's River and its tributaries between Wallacia and its nearest junction with Cowmung River. Most of the original populations seem to have been cleared for pasture or inundated by the Warragamba Dam. Some remnant stands occur along the Nepean River and some good stands occur at Dents Basin and Kedumba Creek within the Warragamba catchment area. The species discussed here does not include the closely related gum which occurs on the northern tablelands of NSW.

Description

This eucalypt is a large white gum which when fully grown is a beautiful tree. It usually has a little persistent flaky bark at the butt in large trees. The buds and fruits are in sevens and quite small. The juvenile leaves are ovate-lanceolate and broader and shorter than *E. viminalis* with which it has some affinity. This species may be mistaken for *E. viminalis* but the flower clusters of *E. benthamii* make it distinct.

Ecology

Camden White Gum is characteristically found on good soils on flats and levee banks adjoining sizeable rivers. The mean maximum temperature of the hottest month is around 26°C and the mean minimum temperature for the coldest month about 4°C with a few light frosts each year. The rainfall is about 1100mm a year with moderate peaks in summer and autumn.

Conservation

It is a species of fertile soils and, in the area in which it occurs, these soils are of quite limited extent. Most of it has been cleared for agricultural development, particularly for pasture, so that the existing stands are mere remnants of the original forests which, in any case, were never very extensive. A good number of the surviving stands were inundated by the Warragamba Dam and the best ones which persist are those within the Warragamba Dam Reserve above the waters of the dam. If these remain undisturbed preservation is adequate, providing regeneration is ensured.

Special Features

Camden White Gum has been recorded from the Dorrigo area of NSW where there is a white gum which is very similar to the species in the Camden area. It is described as *E. benthamii* var. *dorrigoensis* but reexamination suggests it should not be regarded as Camden White Gum although the definitive botanical work remains to be completed. In the mid-1800s, a substantial collection of eucalypts was established on the little peninsula at Cap d'Antibes near Cannes in France at a garden called Villa Thûret. Amongst the fifty or so eucalypts established there and examined by the noted botanist, Naudin, was *E. benthamii*. It is still living and is a large mature white gum over 30m tall and with a stem diameter of over 1.5m. It is somewhat surprising that a species with such a limited distribution should have been amongst these earliest introductions to Europe, but its proximity to Camden is probably the explanation. *E. benthamii* is related to a number of gums in the *E. viminalis* group although it is a distinct species and one of the relatively few which prefer more fertile soils. This preference has made it particularly vulnerable during land clearing and the preservation of the few remaining stands is highly desirable.

12
18
24
30
36
42
111
117
123
129
135
141
147
153
159

Eucalyptus brockwayi C.A. Gardner **Dundas Mahogany** (SIS:C F.T.S. -)

This eucalypt is a rather tall tree of the Norseman area of Western Australia.

Distribution

It is known from a number of small stands in the general vicinity of Norseman. All known localities are close to access tracks which suggests that more intensive survey may disclose other stands in the same general locality.

Description

E. brockwayi is a rather tall single-stemmed tree reaching up to 22m in height. It has shining leaves in bright sunlight. The bark is mostly white with some mottling and the fruit is small and rather clustered, with a marked urceolate form. It is easily confused with *E. floctoniae* which however has larger fruit and the leaves are duller.

Ecology

This species occurs in small stands scattered in the Noresman area often on somewhat rocky slopes. It is not easy to discern from field inspection a correlation between its occurrence and the soil or geological conditions of the site, although it is presumed there must be some such correlation. The mean maximum temperature of the hottest months are quite high about 35°C and there are a few light frosts in winter and the mean minimum temperature of the coldest month is about 5°C. The rainfall is about 300mm with a pronounced winter peak and a very long summer drought.

Conservation

The known stands are apparently on unoccupied crown land and there may be some on the Brockway Forest Reserve south of Norseman although this needs verification. Under present land use conditions it is unlikely that the stands will be disturbed although some are on roadsides which could be reduced by realignment. It is in a zone in which mining activity is frequent so that there is a possible source of some destruction. It seems likely that precise survey would disclose additional stands since most of those now known are near access routes. Although the position is not as urgent as with a number of other eucalypts it would be wise to know the full extent of the occurrence of the species and to set up some reserved areas.

Special Features

It is a fine tree with an extraordinary physiological capacity to grow in quite dry or even arid conditions. It has been quite widely planted as an ornamental tree, both within Australia and overseas and is useful for shelter. Plantings around Broken Hill in western NSW are a good example of this. The shiny leaves are attractive as is the smooth whitish trunk. It is extraordinary that it reaches such a considerable height in an area of low rainfall.

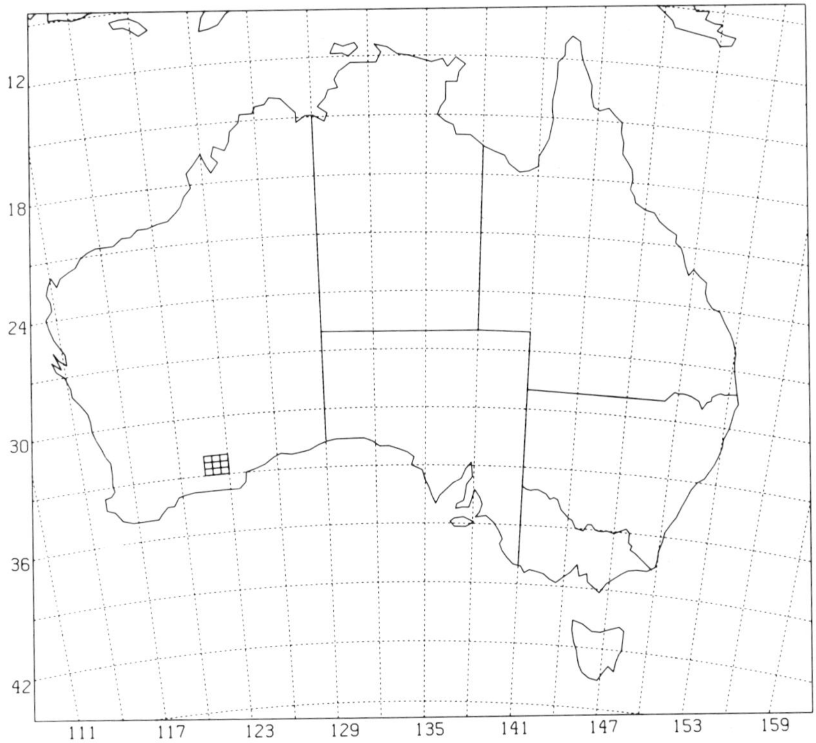
12
18
24
30
36
42
111
117
123
129
135
141
147
153
159

Eucalyptus burdettiana Blakely and Steedman **Burdetts Mallee** (SICBC F.T.S.—)

This is a highly restricted small mallee in Western Australia near Hopetoun.

Distribution

The only site on which this species occurs is on the southeast slopes of east Mt Barren which is in the Fitzgerald River National Park a little west of Hopetoun in Western Australia.

Description

This species is a small shrub of mallee form no more than about 2m in height with smooth stems and rather yellowish branches and somewhat coarse leaves. The fruits are quite large with a thickened dilated peduncle and, when ripe, they are quite woody and up to 3cm in diameter. It is closely related to *E. megacornuta*.

Ecology

Burdetts Mallee occurs on the south coast of Western Australia in an area of very equable climate, similar to that of *E. acies*. The summers are mild with the mean temperature of the hottest month only about 25°C and with virtually no frosts in winter with the mean minimum temperature of the coldest month about 6°C. The rainfall is about 600mm and very reliable with a pronounced winter peak and only light falls through the remainder of the year. It is found on rocky slopes which are well drained and the sites do not have any characteristics which would indicate that they are particularly unusual.

Conservation

The known stands are within the Fitzgerald River National Park which has a management aim of preservation of the natural vegetation. This species is not under threat at present and would be so only if some activity intruded into the park or further development of access lead to disturbance.

Special Features

A species such as this with a very highly restricted distribution, with a distinct form and which is quite taxonomically discrete is typically regarded as relict species in the eucalypt vegetation. There are quite a number of species in this category both in eastern and western Australia. *E. burdettiana*, because of its highly restricted distribution, is prominent in this way. Presumably with higher rainfalls it could have been more widely distributed in the past.

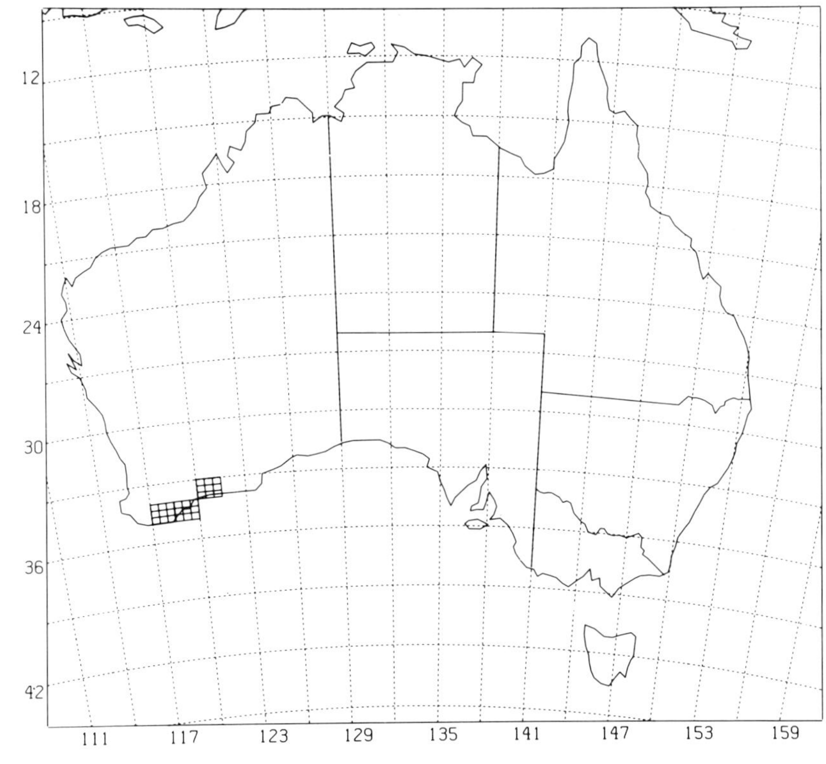
12
18
24
30
36
42
111
117
123
129
135
141
147
153
159

Eucalyptus burgessiana L. Johnson and D. Blaxell **Burgess Mallee** (MAKIF F.T.S.—)

Distribution

This species is known in mallee stands in the Blue Mountains west of Sydney in the vicinity of Faulconbridge.

Description

Burgess Mallee is usually a multi-stemmed robust mallee with coarse leaves somewhat slatey in appearance. The buds and fruits are somewhat larger than is common in *E. stricta* although the latter species is rather more vaiable than many species.

Ecology

This eucalypt occurs on stoney ridges in sandstone area of the Blue Mountains associated with an array of shrubby sclerophyllous species of typical sandstone genera. The mean maximum temperature of the hottest month is about 27°C and the mean minimum of the coldest month is about 5°C.

Conservation

Some loss of stands by suburban expansion has occurred but the main known areas are in the Blue Mountains National Park. The position of this species therefore seems fairly secure.

A closer survey would be likely to reveal additional stands in the same general area.

Special Features

The close relationship to *E. stricta* leads some authorities to feel that it would be better regarded as a subspecies of that species. Nevertheless it is a distinct taxon and merits preservation.

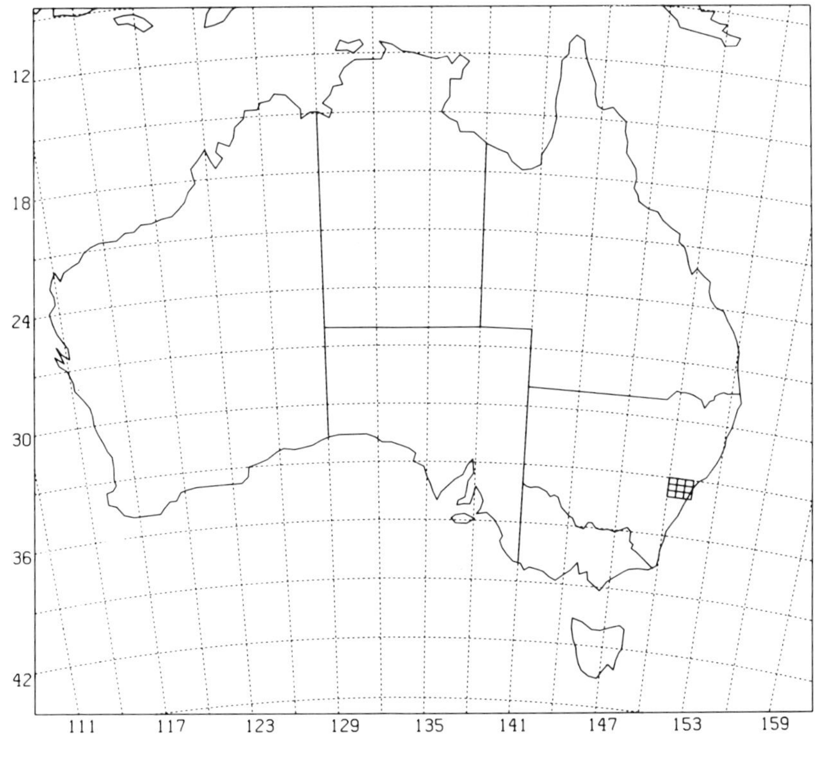
12
18
24
30
36
42
111
117
123
129
135
141
147
153
159

Eucalyptus caesia Benth. **Gungurru** (SIVCG F.T.S.—)

Gungurru is one of the highly valued ornamental small trees amongst the eucalypts.

Distribution

It is known from a number of sites in Western Australia almost always associated with large granitic outcrops between longitudes 117° and 123° and between the latitudes of 31° and 35°S. The populations at all sites are small and they are mostly separated a considerable distance from each other.

Description

E. caesia is a small straggly tree with glaucous leaves, stems and fruits, sometimes making a small tree of 8-10m in height. The flowers are fairly large and when fully open are up to about 4cm across. The flowers are a beautiful shade of pink, although this is variable. The trees most favoured for cultivation as ornamentals are from the Boyagin Rock area which have both the most suitable form and flower colour. The fruits are fairly large, about 2cm diameter or more, and are more or less bell-shaped with a somewhat contracted rim. They are highly glaucous while immature.

Ecology

The tree is almost completely confined to thin soils associated with large granite outcrops. The mean maximum temperature of the hottest month is around 30°C and the mean minimum temperature of the coldest about 5°C with a few frosts each year. The rainfall is generally about 400mm with a pronounced winter peak. There are several months which are rainless during summer.

Conservation

Several of the sites are within nature reserves in particular Mt Caroline, Chiddar Copping Rocks, Billy Cating Hill and Boyagin Rock. In the latter case the population is only partly within a nature reserve. Because of the genetic variation and the very limited nature of the populations, each locality is of importance and merits preservation. Because of the rocky nature of the ground in most sites, there is little risk from land development even where it is outside an established reserve. However part of the population at Boyagin Rock is vulnerable to land clearing for pasture improvement. The ornamental value of the tree is such that there may be risk of some damage by indiscriminate seed collection and this requires some regulation.

Special Features

The considerable ornamental value of the species makes it one of the most prized for growing in gardens where the climate is suitable. Its scattered distribution is consistent with the relict pattern which is so common with rarer species of eucalypts.

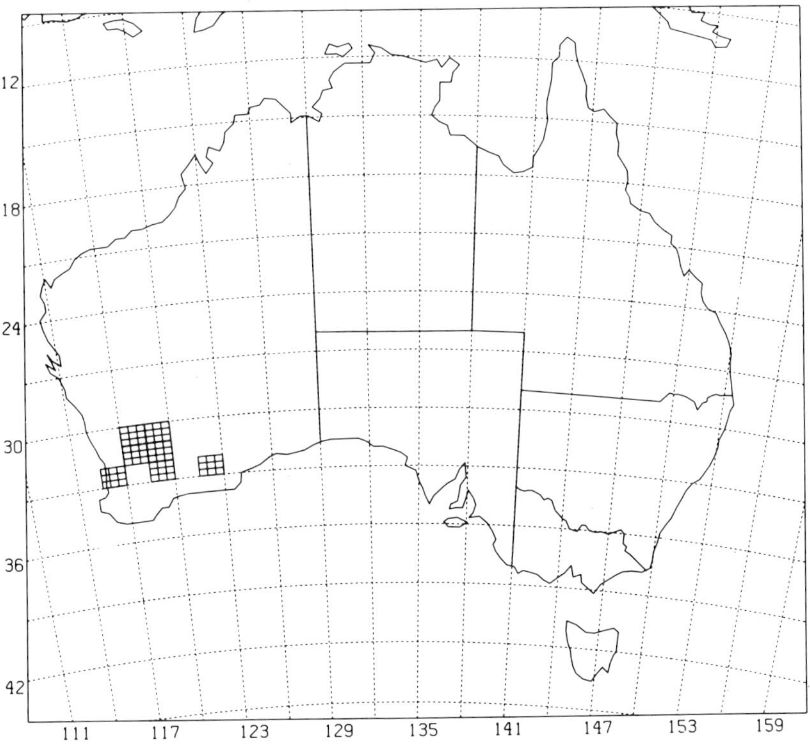
12
18
24
30
36
42
111
117
123
129
135
141
147
153
159

Eucalyptus calcicola M.I.H. Brooker **Hamelin Bay Mallee** (MABEN F.T.S. 154)

Hamelin Bay Mallee is a small shrubby mallee of very restricted distribution in Western Australia.

Distribution

This species is found near Cape Freycinet and at Hamelin Bay near Karridale in Western Austrlaia about 15-25km northwest of Cape Leeuwin. It was described only in 1974. It is quite possible that further search will reveal a few more stands but it is necessarily because of its site preferences, a species of very restricted distribution.

Description

It is a small many-stemmed mallee up to about 3m in height with rather heavy leaves. It has strongly ribbed buds and fruits in clusters of 7.

It is smooth barked with stems seldom more than 7cm in diameter.

Ecology

This mallee occurs on large calcereous dunes in the Hamelin Bay area and nearby parts of Cape Leeuwin in about latitude 34°S in Western Australia. It occurs quite close to the coast at an altitude of less than 100m. The mean maximum temperature of the warmest month is about 24°C and the mean minimum of the coolest around 10°C. There are no frosts on the seaward slopes although very light frosts may occur in the nearby depressions. The rainfall is over 1000mm a year with a winter maximum and some rain through most of the year except during midsummer. It occurs as the dominant mallee on exposed coastal low sclerophyllous scrublands made up of species of many genera. Nearby *E. calophylla* and *E. megacarpa* are found.

Conservation

The Hamelin Bay stand is in an area which will probably be reserved. The nature of the vegetation, the exposed character of the sites, and the sand dunes on which it grows, make it unlikely to be damaged or eliminated. Additional survey might well reveal a few more occurrences since it has not yet been known long enough to have been thoroughly searched for. Reservation of one of the main sites is merited.

Special Features

This eucalypt is of particular interest because of its capacity to grow on calcareous dunes with soils of high pH and is one of the few eucalypts in this climatic zone which is capable of thriving on such sites. The main tree which does well in such soil types is *E. gomphocephala* which is widespread further north.

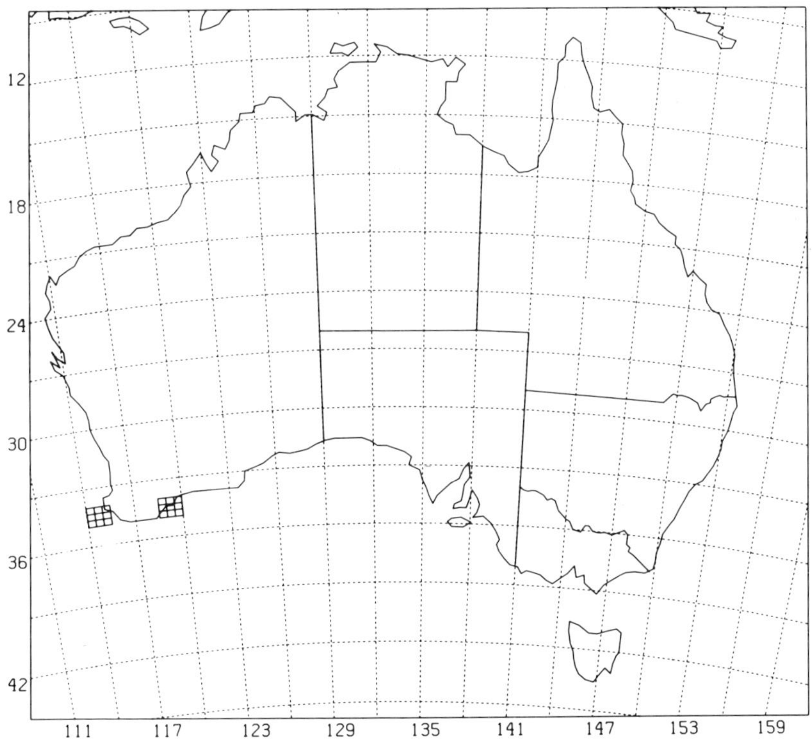
12
18
24
30
36
42
111
117
123
129
135
141
147
153
159

Eucalyptus camfieldii Maiden **Camfields Stringybark** (MAHCE F.T.S. 94)

This eucalypt is a small stringybark found in the vicinity of Sydney.

Distribution

This species is found close to Sydney particularly north of the harbour in localities such as Mosman, Northbridge and Kuringai Chase and also south of Sydney near Waterfall. It no longer exists on a number of sites because of urban development.

Description

This eucalypt is a small tree and at times no more than a mallee. The bigger specimens reach about 10m in height. It is a typical stringybark with closely aggregated fruits, but, amongst the stringybarks, it has very distinctive juvenile leaves some of which are almost orbicular.

Ecology

Camfields Stringybark occurs on very poor coastal soils of a lateritic character at a latitude of about 33.25°S and an elevation of approximately 300m. There are just one or two frosts in the year. The mean maximum temperature for the hottest month is 27°C and the mean minimum for the coldest month about 7°C. Mean annual rainfall is about 1100mm with a late summer maximum. There is usually some rain in each month of the year, although at times it may be subject to substantial drought. It occurs as low depauperate forest with many associated sclerophyllous shrubs and often *E. haemastoma* and *E. oblonga* are found in the vicinity.

Conservation

Urban spread may yet threaten stands of the species although some of them are in national parks or other reserves. Because there is so little remaining, a detailed survey to determine the availability and extent of good stands not so far reserved would be worthwhile.

Special Features

It is of botanical interest because amongst the stringybarks it is unusual in some of its characters particularly the juvenile leaves. Also it is associated with very poor soils on sites which are more or less erosional remnants and its distribution suggests it may be a relict species. The area of total distribution is about 100km north to south so it is a quite restricted endemic species of the Sydney region.

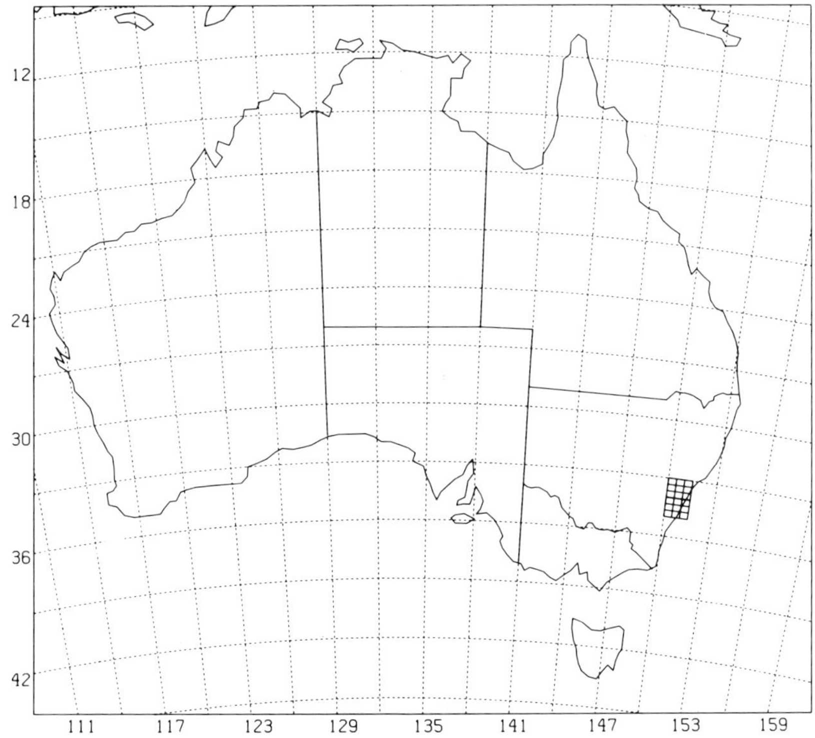
12
18
24
30
36
42
111
117
123
129
135
141
147
153
159

Eucalyptus cneorifolia DC **Kangaroo Island Narrowleaved Mallee** (SIP:K F.T.S. 130)

Commonly a mallee this species can at times on suitable sites become a small tree up to 10m tall.

Distribution

It is found on the eastern half of Kangaroo Island where it was once reasonably widespread. Most occurrences are now remnants from clearing and are along roadsides or on sites which have escaped conversion to pasture. A couple of very small occurrences are known on the mainland near Waitpinga towards Cape Jervis.

Description

It is seen mostly as a dense-crowned tall mallee or a small tree of some 10m in height. It has a rough bark near the butt but the trunk and upper branches are smooth. The adult leaves are quite narrow and shining. In open sites it makes a dense bushy shrub. Buds and fruits are small in clusters of 7-11 and the juvenile leaves are small and narrow.

Ecology

This eucalypt occurs mainly in often quite deep sandy soils and on some sites which are relatively poorly drained. It is found about latitude 36°S and generally not above 100m in altitude and often quite close to the sea. The climate is cool and sub-humid with the mean maximum temperature of the warmest month being about 24°C and the mean minimum for the coolest month about 10°C. Generally there are no frosts. The mean annual rainfall is about 450mm with a winter maximum. It is associated with many sclerophyllous shrubs in numerous genera which in the original undisturbed areas were quite dense. Nearby eucalypts are *E. cosmophylla*, *E. diversifolia* and at times *E. anceps*. The small occurrence on the mainland near Waitpinga does not seem to differ appreciably from the main population on Kangaroo Island.

Conservation

Originally the population of *E. cneorifolia* was quite extensive on the eastern end of Kangaroo Island but most has been cleared and while there are still some specimens along roadsides, it must be considered in rather a vulnerable position unless specific reservations are made. The Waitpinga occurrence is on private property and is well worth preserving because it is a disjunct population. The present owner seems well aware of its interest and value. There are probably some occurrences on Kangaroo Island which are on public land but careful scrutiny and additional survey would be necessary to determine the full extent of surviving stands.

Special Features

This mallee is a restricted South Australian endemic. It is of particular interest because it was one of the species used in earlier days for the production of eucalyptus oil although that industry with this species has long since ceased. It is a tree with considerable ornamental value and it forms dense thickets which are a very effective windbreak. It grows in climates subject to considerable wind and would be useful for planting for shelter. Many of the remnant stands do perform this function.

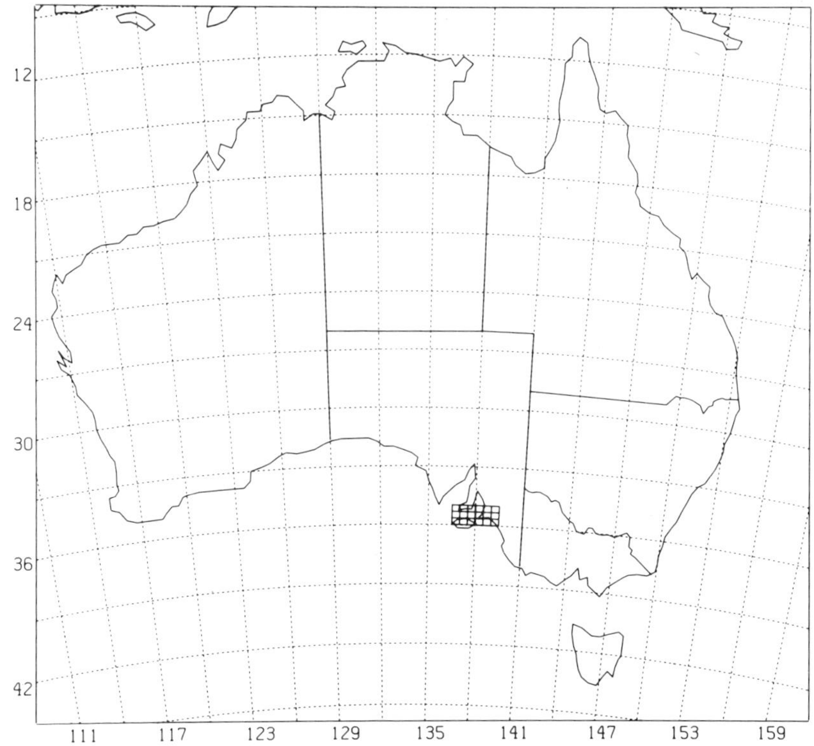
12
18
24
30
36
42
111
117
123
129
135
141
147
153
159

Eucalyptus conglomerata Maiden et Blakely **Swamp Stringybark** (MAHEJ F.T.S. 159)

It is a small crooked tree found in southern Queensland.

Distribution

There have been few records of the species, the type locality of which is near Beerwah. It is also reported near Coolum and from between Beerwah and Caloundra.

Description

This species is a small stringybark at times reaching 12m but often only as a mallee. It is a crooked straggly tree. The bark is not quite typical of most stringybarks in that it has irregular horizontal fissures as well as longitudinal fissures so the ends of the outer layers may be raised from the stem. It is distinct from other stringybarks with pointed buds and small fruits, by having depressed discs and sunken valves whereas in other species these are more or less level or somewhat raised.

Ecology

Swamp Stringybark grows on coastal flats or slight depressions in gently undulating areas which are wet and swampy during the wet season. They are deep podzols of very low fertility. The mean maximum temperature of the hottest month is about 30°C, and the mean minimum of the coolest about 10°C. There are scarcely any frosts. The mean annual rainfall is 1500mm with a summer peak but with a fairly good distribution through from April to November.

Conservation

Inadequate knowledge of the distribution of the species makes it difficult to be sure of the extent to which it is under threat. It is reported on a forest reserve at Beerwah but there seem to be very few specimens in that area. In spite of the very low fertility of the sites on which it occurs, agricultural development of these types of sites are now possible and it seems likely that there is a distinct threat to the species. Setting up of a suitable reserve seems desirable.

Special Features

Swamp Stringybark is interesting because of its adaptation to very poor swampy sites. These are conditions under which very few eucalypts grow.

12
18
24
30
36
42
111
117
123
129
135
141
147
153
159

Eucalyptus cordata Labill. **Heart-leaved Silver Gum** (SPINO F.T.S. 53)

This species is a small tree endemic to southeastern Tasmania.

Distribution

There are a number of localities within a radius of 70 miles of Hobart from which the species has been reported. There are sites at Moogana, Mt Lloyd, Chimney Pot Hill, the Tasman Peninsula and Bruny Island.

Description

This eucalypt varies greatly from a low straggly shrub of 2 or 3m to a straight tree of about 20m. Under natural conditions the juvenile form of leaf is usually persistent in the adult tree and it is highly glaucous. The buds and fruits are in clusters of 3 and they also are glaucous.

Ecology

Heart-leaved Silver Gum occurs on a variety of sites, all of very limited extent, around latitude 43°S and at altitudes of from about 150-700m. The mean maximum temperature of the warmest month is about 21°C and the mean minimum of the coldest month about 1°C. There are up to about 50 frosts a year and snow always occurs at the higher altitudes. Mean annual rainfall covers a fairly wide range of from about 500 to over 1000mm according to altitude. It occurs on a variety of sites and sometimes in small depressions which are often marshy. Otherwise it is on moderately steep slopes in wet sclerophyll forests. The soils are mainly yellow podzolics and the parent rock is predominantly dolerite. Other eucalypts nearby are frequently *E. delegatensis, E. regnans* and *E. obliqua* and, on the wetter sites, *Leptospermum* and sedges are present.

Conservation

The occurrence of the species in small local patches scattered over a substantial area make it difficult to know the full extent of all the stands. Some are threatened and have already been damaged by urban development and others are on private land where they survive only casually in relation to current land use. The tall forms in the Mt Lloyd area certainly merit preservation as well as the dwarf more shrubby forms elsewhere. More detailed survey is a necessary starting point to prepare a conservation plan.

Special Features

This was one of the eucalypts first described. Labillardiere published the description in 1806. Although it occurs in areas which are not especially cold in winter it has proved a reasonably frost-hardy species and has been introduced into places in Europe. Like a number of other species with this type of distribution it appears to be a relict species now confined to a few scattered sites. Here, the tree can survive without being overtaken by the competition of more vigorously growing eucalypt species.

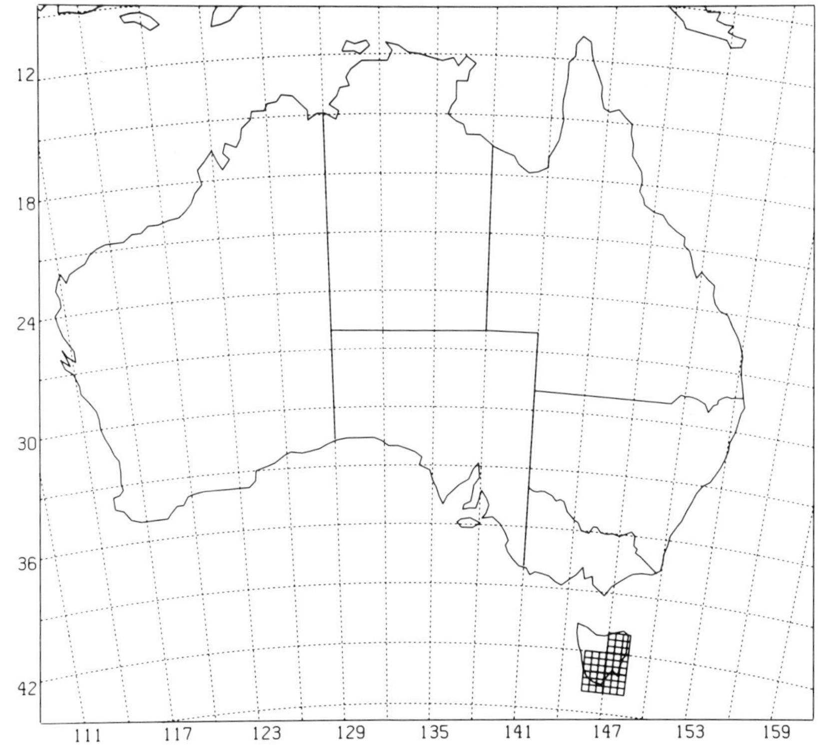
12
18
24
30
36
42
111
117
123
129
135
141
147
153
159

Eucalyptus coronata Gardner **Crowned Mallee** (MAA:C F.T.S.—)

E. coronata is a small mallee from east Mt Barren near Hopetoun, Western Australia.

Distribution

It has a highly restricted distribution on the southeastern slopes of east Mt Barren and one or two other localities nearby. All locations are within the Fitzgerald River National Park.

Description

It is a small mallee up to about 3m in height and rather heavy leaves. The fruits are up to 3 cm in diameter with strong protuberances on the disc and from the valves. The leaves are rather heavy-textured.

Ecology

This eucalypt occurs at a few sites on the southeast slopes of east Mt Barren one of which also supports *E. burdettiana.* The climate there is mild with the mean maximum temperature of the hottest month being about 20°C and the mean minimum temperature of the coldest month about 5°C. There are virtually no frosts. Rainfall is a reliable 600mm per annum with a winter peak. It occurs in shrub heath about 1-2m tall with many shrub species from a wide range of genera including *Banksia, Hakea* and *Acacia.* The soils are well drained, somewhat stony and derived from gneiss.

Conservation

Although highly restricted in distribution all natural stands are in the Fitzgerald River National Park a management aim of which is the preservation of natural vegetation. If management takes into account the need for regeneration of the species it is not at risk of extinction.

Special Features

A distinct species with only *E. aquilina* in a somewhat similar site at Cape Le Grande National Park some 120km to the east, as a close relative. This species is highly restricted in known natural occurrences and is quite distinct which suggests that it is a relict species from another climatic period.

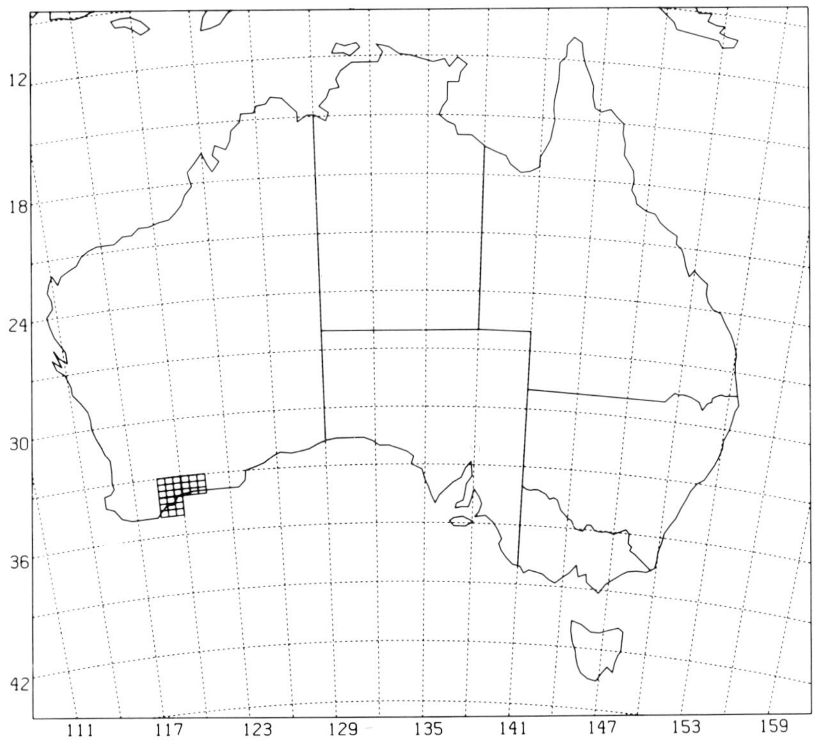
12
18
24
30
36
42
111
117
123
129
135
141
147
153
159

Eucalyptus crenulata Blakely et de Beuzeville **Victorian Silver Gum** (SPICA F.T.S. 87)

E. crenulata is a small scrambling glaucous tree with considerable ornamental value. It is a quite unusual species and there are now probably more plants in cultivation than in the wild.

Distribution

It occurs only in Victoria and in two separate main areas of quite limited extent. One is near Buxton near the Acheron River and the other is at Yering on the upper Yarra. A few other sites have been recorded but a recent inspection failed to find the species at these sites. It has a very restricted natural distribution and a number of sites have been cleared.

Description

This species is a small tree of irregular form, at times reaching perhaps 12m in height, and 25cm stem diameter. It retains the juvenile type foliage throughout its life, the leaves being more or less opposite and sessile. It is always glaucous on the stems and leaves and the leaf margins are crenulate which is a distinctive feature of the species. The buds occur in clusters, often of 11 or more and are highly glaucous with a pointed operculum. These characters together make it a very unusual species without any close relatives.

Ecology

It occurs on humid sites which are sometimes swampy and the Buxton area has sphagnum moss present. This site is quite swampy for much of the year. *E. crenulata* tends to be overtopped by other species, particularly E. ovata, at the edges of the stands.

The summers are comparatively mild and the winters cool with around 25 frosts a year. The mean maximum temperature of the hottest month is abour 25°C and the mean minimum of the coolest month about 2°C. The mean annual rainfall is about 1000mm and is well distributed throughout the year.

Conservation

The Buxton site was included in a small reserve arranged by the Victorian Forest Commission in 1972 and specifically set aside to preserve a natural stand of the species. It would be very desirable to ensure the preservation of the stand at Yering where there is intrusion of weeds and grazing is still evident.

Special Features

E. crenulata is a particularly interesting species for several reasons. It is distinct from all other eucalypts although it has some affinity with *E. bridgesiana*. It occurs on very wet sites which are even swampy and is therefore quite unusual amongst the eucalypts. It is a small species which tends to be overtopped by other species in the natural stands. At the Yering site it is in hybrid combination with *E. ovata* but this does not seem to have occurred at the Buxton site although *E. ovata* is present. It is one of the small number of eucalypt species of limited extent which appear to have developed in relation to cool moist pluvial conditions and which are now restricted to very limited sites. There has been clearing of the original stands which were never very extensive, and all that now persists are very small remnants of those stands. The existing reservation is sufficient to ensure the species survival in a natural stand but it would be very desirable to reserve the Yering stand and any other sites that can be located. An intensive survey for the species is also warranted.

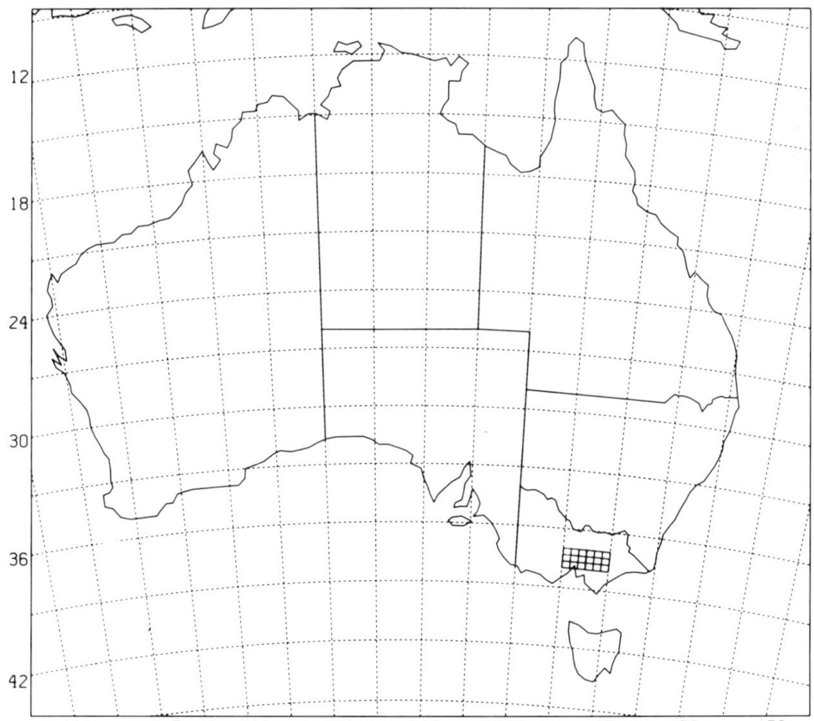
12
18
24
30
36
42
111
117
123
129
135
141
147
153
159

Eucalyptus curtisii Blakely et C.T. White **Plunkett Mallee** (GAA:A F.T.S. 157)

It is a small mallee of southeast Queensland.

Distribution

The main known stands are near Plunkett south of Brisbane, which is the type locality, and another occurrence at Dinmore near the western side of Brisbane. There are a few other scattered locations near Chinchilla and near Beerwah. While spread over a distance of some 200km the stands are all small and scattered.

Description

This is a many-stemmed mallee reaching perhaps 5m in height with rather bright green leaves. The infloresence is more or less terminal with the buds and fruits grouped in clusters of 3 or 7. It is one of the few eucalypts in which the sepals do not form an outer operculum and small calyx teeth persist on the fruit.

Ecology

This species occurs over a considerable area in southeastern Queensland in very small, widely scattered stands. It is often found on rocky soil derived from sandstone or on poor sandy podzols which may be partly waterlogged during the wet season. It is also found on other soil types where the soil is heavier but rather poorly drained. The climate varies in different parts of the range of the species from the warm humid sites in the southeast to drier sites inland. The mean maximum temperature of the hottest month is about 29°C and the mean minimum for the coolest month about 9°C. There are no frosts on most sites although an occasional light frost may occur on the inland sites. The mean annual rainfall is about 1500mm but with less on inland sites. There is a marked summer peak. It grows near other eucalypts particularly *E. maculata* and *E. trachyphloia* and also sometimes near *Callitris endlicheri*. It sometimes tends to form an understorey to other eucalypt species.

Conservation

Although most stands are quite small they are numerous in many scattered localities. Since many sites are rocky they have not been sought for agriculture. Despite the fact that the type locality has been affected by powerline construction the species is not likely to be eliminated at this site. On the other hand, several of the known sites immediately around Brisbane such as Mt Gravatt and Prior's Pocket no longer carry the species and it seems that further urban spread is likely to eliminate some of the surviving stands. There is a small stand on the forest reserve at Beerwah and the western patches still persist. Because of its restricted occurrence there is a potential threat to the species rather than an immediate one. The type locality is not yet a reserve but the interest in the species and the fact that this is one of the best stands of it would merit preservation of that site. The wide geographic distribution of the species and the different conditions under which it goes does suggest that it would be desirable that some of the inland sites and those near Beerwah be preserved.

Special Features

E. curtisii along with *E. tenuipes* form a small distinct group of eucalypts without close natural affinity to other species. They are regarded by Pryor & Johnson (1971) as meriting the status of sub-genus. They are very unusual in their floral structure in that there is no outer operculum and the sepals remain free. The foliage is quite bright green and it is an attractive shrubby tree which would be useful in landscaping.

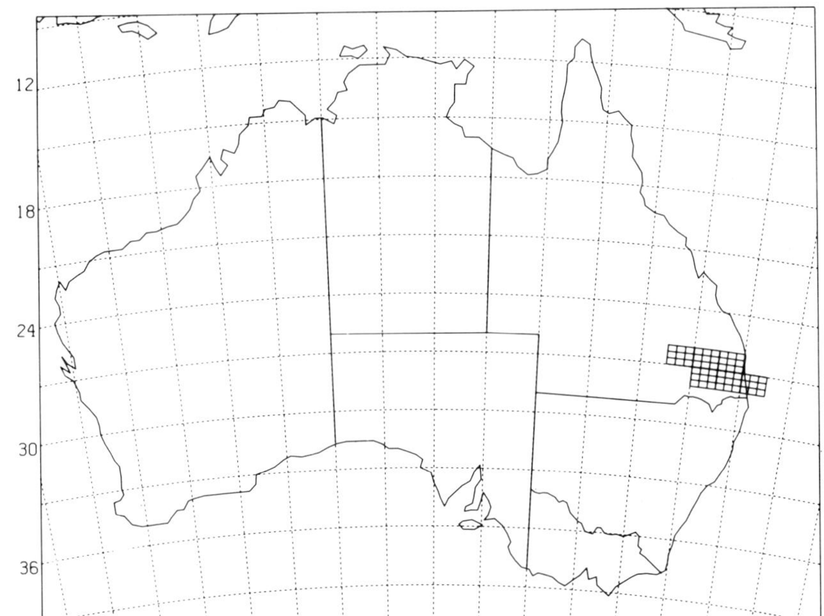
12
18
24
30
36
42
111
117
123
129
135
141
147
153
159

Eucalyptus dendromorpha Johnson et Blaxell **Budawang Ash** (MAKID F.T.S. 216)

It is a medium sized tree of central eastern New South Wales south of Sydney.

Distribution

Budawang Ash is recorded from a number of sites from the vicinity of Sugarloaf Mountain, Fitzroy Falls, Carrington Falls and Budawang Peak. It occurs in stands which are not extensive and is associated with rocky sites.

Description

This ash is related to *E. fraxinoides* and *E. triflora*. It also has some affinity with *E. stricta* and in the smaller forms has been confused with this species. The somewhat pointed buds make it distinct from its nearest related species.

Ecology

Budawang Ash occurs on most soils derived from sandstone usually in the vicinity of rocky outcrops. The mean maximum temperature of the hottest month is about 27°C and the mean minimum temperature for the coolest month about 3°C. Mean annual rainfall is rather high at about 1500mm with some frosts depending on the altitude of the site. It is often found in association with *E. sieberi, E. gummifera, E. stricta, E. smithii* and other species forming a fairly tall forest.

Conservation

This species is in a number of reserves such as Barren Grounds Nature Reserve and Mt Budawang National Park and on other sites such as Sugarloaf Mountain. It is unlikely that any existing form of land use would disturb it because of the nature of the sites.

Special Features

The variation in size of the tree has led to some confusion in the past with *E. stricta* which is a variable mallee species. Some morphological details raised by Brooker imply that it is a distinct species, although somewhat variable in stature according to the site on which it occurs. The restricted nature of the sites around rocky edges of scarps and its fragmented distribution in small stands is like that of several other species which is consistent with the suggestion that it is a relict species.

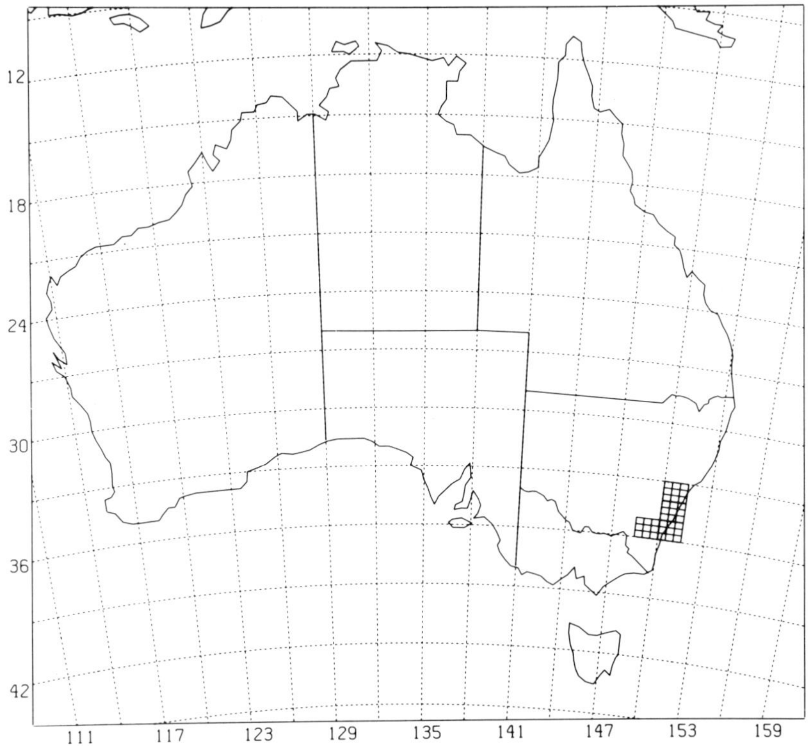
12
18
24
30
36
42
111
117
123
129
135
141
147
153
159

Eucalyptus desmondensis Maiden et Blakely **Desmond Mallee** (SIGCA F.T.S.—)

Desmond Mallee is known only from a limited area of Western Australia near Ravensthorpe.

Distribution

It is found on elevated ground in the vicinity of Mt Desmond between Ravensthorpe and Kundip in Western Australia. It occurs in small open stands scattered through the area.

Description

This is a slender glaucous mallee up to about 4m tall with flexuose drooping branches, glaucous stems and smooth bark. The buds are cylindrical, almost sessile and the fruit is quite glaucous.

Ecology

This mallee occurs on the slopes of Mt Desmond principally in an area with sandy soils and heath vegetation. The mean maximum of the hottest month is about 25°C and the mean minimum of the coolest month about 5°C. It stands well above associated heathy shrubs which are the common sclerophyll genera in that part of Western Australia.

Conservation

Mt Desmond is a locality for mining operations but otherwise there is no evidence of threat to the persistence of the species. It is unlikely that the site will be developed for agricultural use. The species is highly restricted in its distribution, although survey in the area is inadequate because of limited access. Preservation of an adequate sized stand of the species is desirable although the status of the known sites has not yet been determined.

Special Features

E. desmondensis is a rather distinct species with broad relationship to *E. accedens, E. wandoo,* and related species. Like many of the highly restricted endemics of Western Australia it is difficult to determine, by simple site inspection, any feature of the habitat which is peculiar and which would suggest association of the species with those particular conditions.

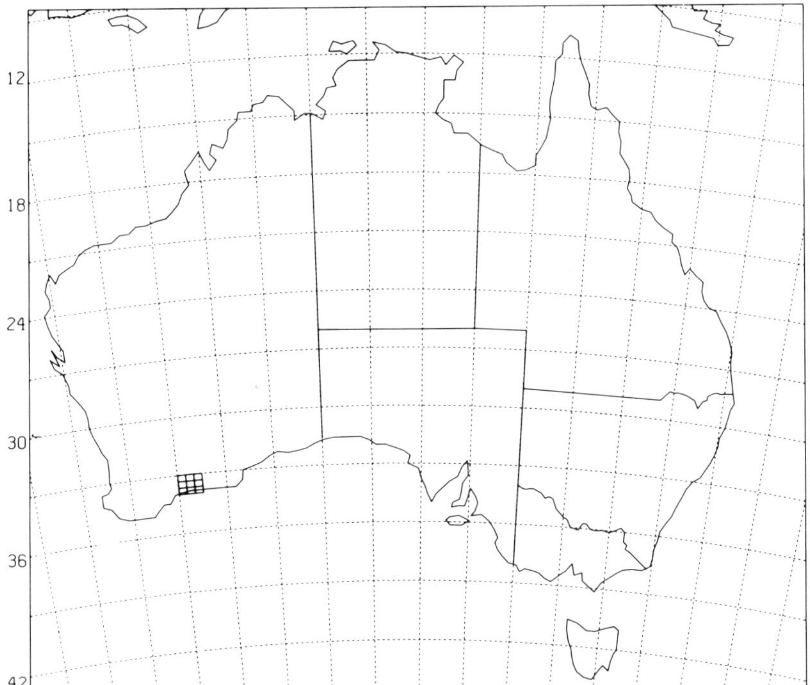
12
18
24
30
36
42
111
117
123
129
135
141
147
153
159

Eucalyptus ficifolia F. Muell. **Red Flowering Gum** (CAFOA F.T.S.—)

Western Australian Red Flowering Gum is one of the most striking of eucalypts when in blossom. It occurs in the southwest of Western Australia.

Distribution

This species has a limited distribution in the southwest of Western Australia generally in the vicinity of Nornalup, Denmark and the Franklin River. The stands are small and scattered.

Description

This species is a distinct bloodwood with the typical rough bark of the group although at times the bark can be fibrous. The leaves are stiff and markedly discolorous. The rather large fruits have a contracted orifice. It has especially brilliant red, orange or crimson flowers in large terminal clusters. In full flower it is a very showy species indeed. It is a small umbrageous tree up to about 10-15m in height.

Ecology

This eucalypt occurs in patches through a forested zone near the south coast of Western Australia. The summers are mild with the mean maximum temperature of the hottest month being about 23°C and the mean minimum temperature of the coldest month about 5°C. The rainfall is quite high being about 1500mm per annum. The soils are sandy with very low fertility which is typical of that part of Western Australia. It occurs near tall forest species such as *E. calophylla, E. diversicolor, E. jacksonii* and *E. marginata.* It has been heavily burnt by wildfire over long periods of time.

Conservation

Most of the existing population is within the Walpole-Nornalup National Park and is therefore well preserved.

Special Features

The flower of this *Eucalyptus* is one of the best known and because of this it has been widely planted both within and outside Australia. It is also a very good shade tree with moderate dimensions and has a dense spreading crown. It is best suited to areas with a winter rainfall and does not thrive in summer rainfall areas. It is one of the three bloodwood species endemic to southwest Western Australia, the nearest relatives of which are in Northern Territory or southeastern Australia. It is part of the evidence for the long term separation of the eastern and western parts of the Australian flora.

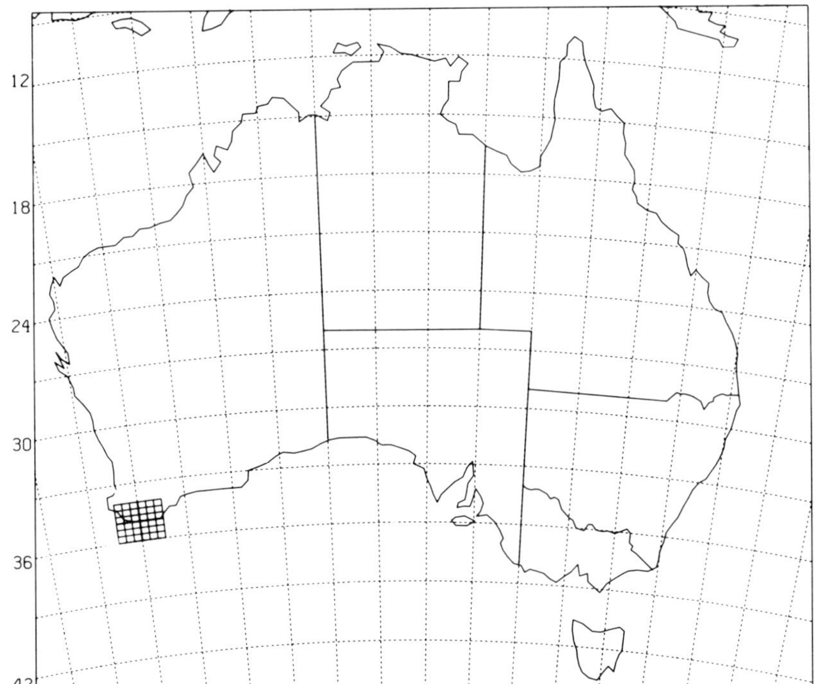
12
18
24
30
36
42
111
117
123
129
135
141
147
153
159

Eucalyptus froggattii Blakely **Kamarooka Mallee** (SUNEF F.T.S. 80)

Kamarooka Mallee is found in a very limited area in about latitude 37°S in the vicinity of Wedderburn and Bendigo in Victoria.

Distribution

This species is found in a few areas of very limited distribution in central Victoria mainly in the Kamarooka district. This area has been farmed and the eucalypt population has been much reduced.

Description

The adult leaves are rather coriaceous with oblique veins. There is some rough box-like bark persisting on the lower part of the trunk. Kamarooka Mallee is a mallee or small tree up to about 10m in height and in the mallee form may have several stems coming from a common rootstock.

Ecology

It is related to other box species such as *E. odorata* from which it differs particularly in the quadrangular buds and fruits. This mallee occurs at an altitude of about 250m in a zone in which the mean maximum temperature of the hottest month is about 30°C and the mean minimum of the coldest month around 3°C. There are a few frosts each year and a period of very hot weather can occur in summer. The rainfall is about 450mm per annum with a winter maximum. This species is found on gently undulating land with well defined drainage channels. The soils are generally red-brown earths sometimes with lateritic gravel.

Conservation

The occurrences of the species have now been very much reduced by farming operations and it is found mainly on privately owned land, roadsides and a small amount on crown land. Because of its original limited distribution and the fact that it has been very much reduced, it is one of the few eucalypt species that is highly endangered and close to extinction. It has been the subject of detailed study by botanists in Victoria which should produce information allowing firm proposals to ensure its survival. It will be important to ensure that regeneration can occur and that some populations are preserved from further clearing by whatever means this is best achieved.

Special Features

Kamarooka Mallee is of special interest because of its very restricted distribution. It is clearly related to other boxes although it is distinct from those that occur in the same area. As with many species of limited distribution, it is difficult to see by inspection, the conditions of the site which make it different enough to carry a distinct species.

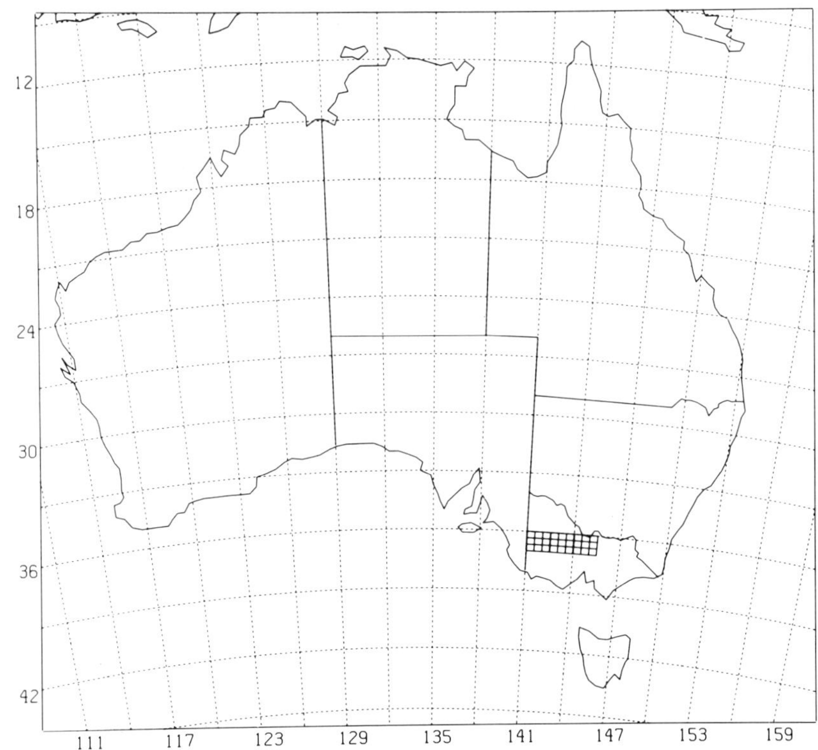
12
18
24
30
36
42
111
117
123
129
135
141
147
153
159

Eucalyptus gregsoniana L. Johnson and D. Blaxell **Wolgan Snow Gum** (MAKHF F.T.S. 167)

Distribution

E. gregsoniana occurs in small patches at relatively high altitudes. It is found near the coastal scarp of the NSW tablelands, west and south of Sydney. It occurs from about latitude 33-35°S and at altitudes of about 1000 metres. The stands are small and with a fragmented distribution.

Description

A small many-stemmed mallee often only up to 2m in height, but occasionally up to 5m. It differs from forms of *E. pauciflora* and related species such as *E. debeuzevillei* in that is has the mallee form. The leaves are narrower than most snow gums and the fruits are relatively small.

Ecology

This species occurs in the cool humid zones where the mean maximum temperature of the hottest month is about 24°C and the mean minimum temperature of the coolest month is about 0°C. There are probably up to about 50 frosts a year and the rainfall is about 1200mm per annum with a moderate summer maximum. There are light falls of snow in most years. The plant occurs on skeletal soils of low fertility derived from sandstone and other siliceous rock. It is closely limited to soil types derived from certain geological substrates. Other eucalypts sometimes found with it are *E. stricta, E. sclerophylla, E. dives, E. sieberi* and *E. mannifera.* A number of small sclerophyll shrubs form a shrub community found in association with this species.

Conservation

It occurs on numerous sites from the vicinity of Mongarlowe to north of Newnes, N.S.W. The nature of sites is not likely to lead to agricultural or forestry use. There are also many occurrences so that the more remote possibilities of engineering structures eliminating large amounts of the whole population are quite small. Also, some sites are in reserves such as Budawang National Park. The threat to the elimination of the species is therefore not very great.

Special Features

The relationship to Snow Gum, *E. pauciflora,* is a matter of biological interest as it provides an example of a form difference with some intermediate gradations. The differences have been recognised by botanists as sufficient to constitute a separate species. Whatever the rank of the taxon, which some would regard as a subspecies, it nevertheless is a distinctive form which merits preservation.

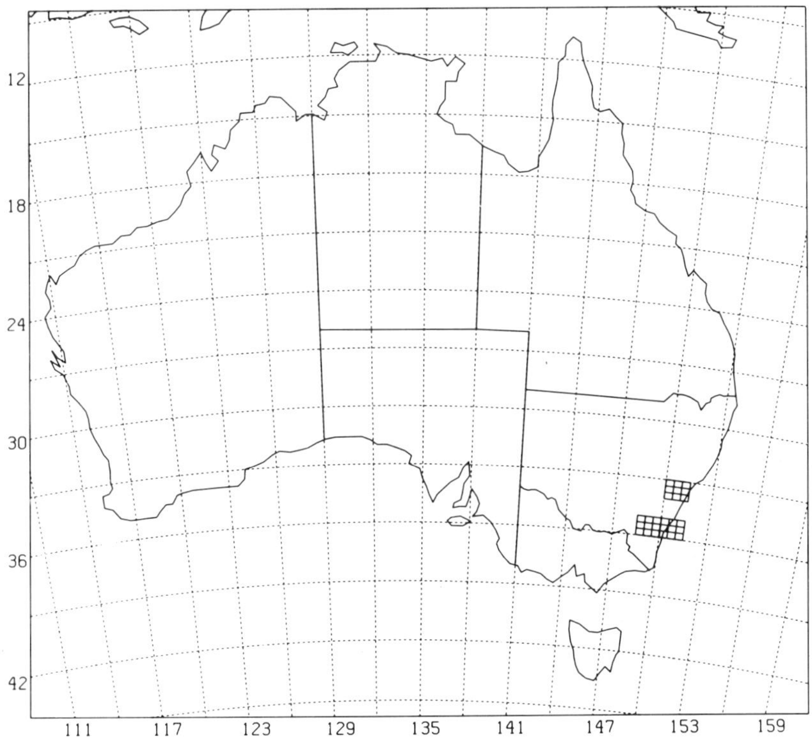
12
18
24
30
36
42
111
117
123
129
135
141
147
153
159

Eucalyptus insularis Brooker **Twin Peak Island Mallee** (MADEM F.T.S.—)

Distribution

The only known occurrences of *E. insularis* are a site in Cape Le Grande National Park east of Esperance in Western Australia and on north Twin Peak Island in the Recherche Archipelago. In both cases the occurrences are quite limited.

Description

This species is a small mallee up to about 2.25m in height. It has narrow rather bright green leaves, yellowish stems and small slender clusters of stems in each clump. It is a highly distinctive species of the Monocalyptus group of Western Australia.

Ecology

This mallee occurs in rocky clefts on granitic sites in about latitude 34°S. The climate is mild with the mean maximum temperature of the hottest month about 23°C and the mean minimum of the coldest month about 5°C. The highly reliable rainfall is about 600mm per annum with a pronounced winter peak. The skeletal soils are well drained and derived from massive granite.

Conservation

Both known occurrences of this mallee are in a national park and therefore well protected and likely to be preserved indefinitely.

Special Features

This species was recognised only quite recently although a few specimens had been collected a decade or two earlier. It is a very distinctive species which has a highly limited distribution. The two known occurrences together making it one of the most limited of all eucalypt species and comparable in this regard with *E. burdettiana* and *E. coronata* in Western Australia, and *E. pumila* in eastern Australia. The habitat requirements of the species suggest it is a relict survivor from a time of more widespread distribution of the species in climatic conditions different from those at present. The occurrence of the species on both the off-shore island and the mainland suggests that it may have had continuous distribution during periods of lower sea level.

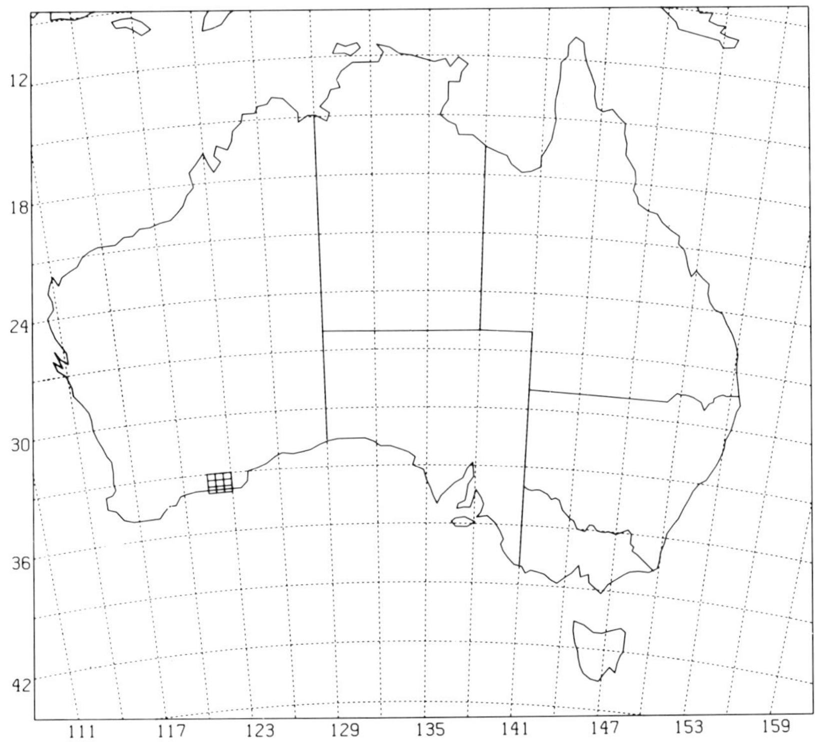
12
18
24
30
36
42
111
117
123
129
135
141
147
153
159

Eucalyptus kartzoffiana L. Johnson **Araluen Gum** (SPINB F.T.S.—)

A good sized tree of limited distribution found near Araluen in New south Wales.

Distribution

This species is found in valleys and higher slopes from just a little upstream from Araluen up to the tableland towards Braidwood and Major's Creek. It is also found in one or two adjoining gullies in the Araluen Valley.

Description

This species is a fairly tall gum of about 30m in height with some persistent rough bark towards the base. It has markedly glaucous juvenile leaves and glaucous stems with some glaucousness on the adult leaves. The fruits are small, in clusters of 7 and the upper stems are smooth with decorticating bark.

Ecology

This eucalypt occurs in areas with a mean maximum temperature of the hottest month around 28°C and the mean minimum temperature of the coldest month about 3°C. The rainfall is about 900mm per annum with a distinct summer peak. It occurs on riverbank soils and along drainage depressions. The sites are well drained and range over altitudes from the valley bottoms to the edge of the tableland. Other species occurring in the vicinity of this species are *E. viminalis, E. maidenii, E. angophoroides* and some *E. globoidea* together with a few tall shrubs, especially *Acacia dealbata.*

Conservation

A large proportion of the orginal stands of this species has been cleared for pasture for grazing and to some extent for orcharding and other cultivation. The stands now to be found are remnants and most are on private property or roadside verges. The species regenerates vigorously, given the opportunity by exclusion of grazing. There are a few restricted small patches where these conditions exist and young plants can be found. The species is found along creeks into steep areas along the edge of the tablelands and there may be more on Crown land or state forest. A stand of about 10ha towards the headwaters of Araluen Creek is in uncleared forest which is probably on private property. It is highly desirable that some stands be retained in a reserve.

Special Features

This is a remarkably vigorous tree which may well have economic potential as well as being of ornamental value. It can become quite a large tree although most of the big specimens seem to have been cleared. It is clearly related to species such as *E. macarthurii* and *E. viminalis* although the juvenile leaves, which are highly glaucous and rather ovate, separate it very distinctly from those species. It is a species which, in spite of these relationships, is quite distinctive and yet of very restricted occurrence. The sites on which it occurs are like others which occur in rough dissected areas to the south in the Tuross River area and it seems possible that other occurrences of the species might be found.

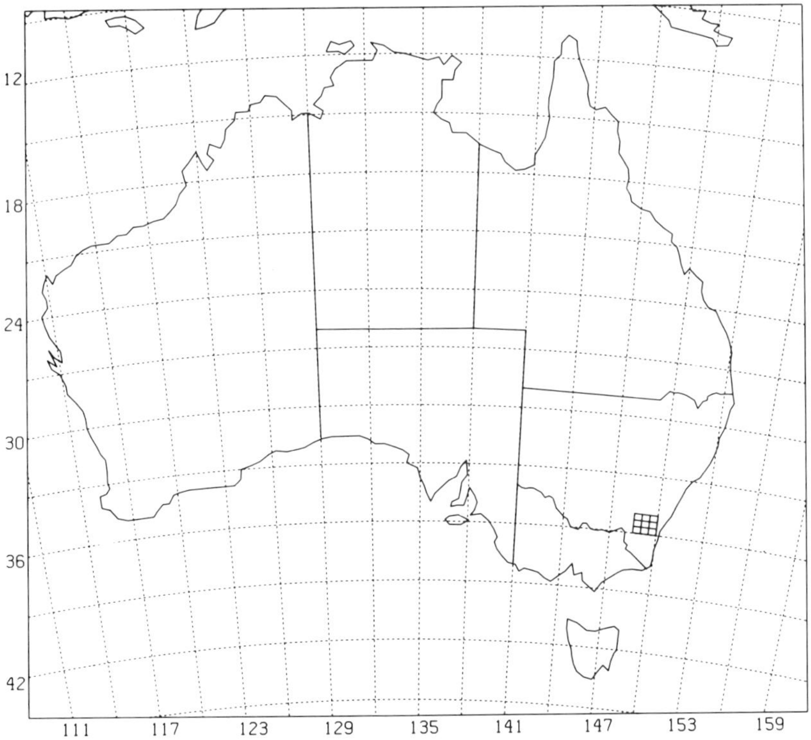
12
18
24
30
36
42
111
117
123
129
135
141
147
153
159

Eucalyptus kitsoniana Luehm. Maiden **Gippsland Mallee** (SPIAC F.T.S. 78)

A good sized mallee or small tree found on either side of Melbourne in Victoria.

Distribution

Gippsland Mallee occurs near the coast in Victoria in the Gippsland district south and east of Melbourne, in the vicinity of the Otways to the west. It is usually within 30-40km of the coast in about latitude 38°S and at elevations generally below 100m. It extends over a distance of about 500km.

Description

It is a moderately tall mallee up to 10m in height and usually with 3 or 4 stems. Has rather broad juvenile leaves, opposite, sessile, green and rather heavy-textured large adult leaves. The buds and fruits are tightly clustered in groups of seven. On favourable sites some specimens are taller than average and on some sites it may be a single-stemmed tree.

Ecology

It grows on coastal plains and small flats and is often associated with streams, acid soils and relatively poor surface drainage. The mean maximum temperature of the hottest month is about 23°C and the mean minimum of the coolest is about 5°C. There are few frosts at the inland sites and none nearer the coast. The mean annual rainfall is about 1000mm which is well distributed but with a moderate winter maximum.

Conservation

Many of the original stands of the species have been cleared for farming but it persists in areas where the conditions are more swampy. Many of the sites are on roadside verges. Detailed survey may disclose that some reasonable stands of the species still persist but none have been seen. it seems likely that there are some stands on the northeast of Cape Otway on state forest or crown land which should be relatively easy to protect. There are some small occurrences in Wilson's Promontory National Park.

Special Features

E. kitsoniana is moderately widely distributed in a narrow strip on wet, cool sites. It is considered to be related to *E. neglecta* and perhaps more remotely to *E. crenulata* and *E. parvifolia* which have rather similar site requirements. There are no close relatives in Tasmania and the species itself has not been found in Tasmania.

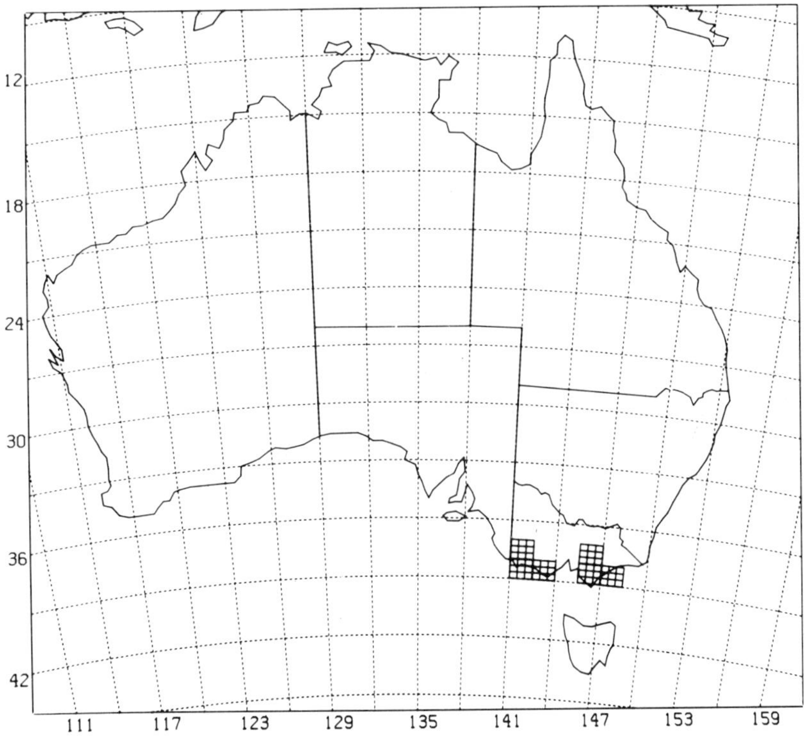
12
18
24
30
36
42
111
117
123
129
135
141
147
153
159

Eucalyptus kruseana F. Muell. **Kruse's Mallee** (SIM:A F.T.S.—)

A small shrub or mallee of ornamental value found in the Goldfields area of Western Australia.

Distribution

It is known from a few locations east and south of Kalgoorlie, particularly at Karonie and Binyarinyinna. It is associated with rocky outcrops near Karonie and at two or three other localities in the general vicinity. The isolated occurrence south of Kalgoorlie at Binyarinyinna is in an area where rock is near the surface but it could scarcely be described as outcropping. The stands are all small with relatively small populations which are well isolated from one another. The areas in which the species occurs lack easy access and a detailed survey would require special arrangements. It is likely that more stands could be located by intensive survey.

Description

It is a small tree or mallee with twisted branches making a rather compact shrubby mallee or a round-headed tree. It retains the juvenile type foliage throughout its life making it particularly distinctive. The leaves are round, opposite and closely set together. The buds are in clusters of 7, highly glaucous and the fruits are also very glaucous.

Ecology

E. kruseana occurs in latitude about 32°S and is associated with granitic soils. The mean monthly maximum temperature is about 30°C and the mean minimum of the coldest month is about 3°C. Frosts occur on 10 or 15 days each year. The rainfall is 200mm per annum with a marked winter peak and a very dry summer. The soils are often skeletal. Other eucalypt species growing nearby are *E. loxophleba*, *E. concinna* and the undescribed *E.* "SUDAE" and a number of tall shrubs such as *Acacia* and *Eremophila*.

Conservation

It is unlikely that agricultural land use will affect the existence of this species. At Karonie, quarrying work associated with the east-west railway has already eliminated some of one stand by the removal of part of the granite hill. A major portion of this stand could be eliminated if quarrying continues as in the past. At Binyarinyinna, road development could reduce the population. Works of this kind could affect other stands from time to time but there is no immediate threat to the survival of the species. However management strategies developed to ensure the long term survival of *E. kruseana* are needed. For example protection of part of the Karonie site may be feasible.

Special Features

This species is one of the few that retain juvenile type foliage right through their adult life. It is valuable for horticultural use and has been planted to some extent in suitable climates in southern Australia. The leaf form, together with the highly glaucous appearance make it valuable this way. The pale yellow flowers add to the attractive appearance of the species.

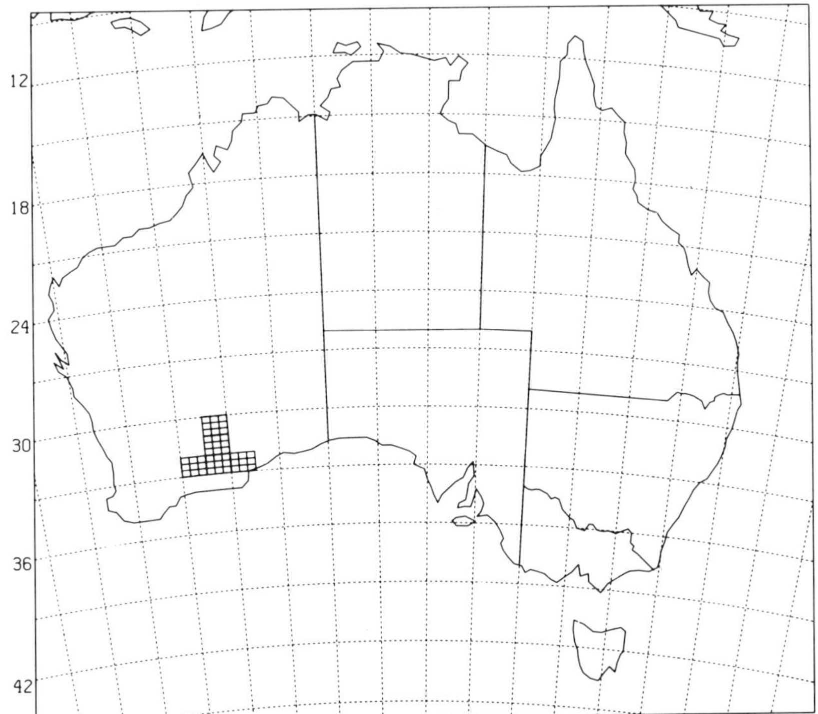
12
18
24
30
36
42
111
117
123
129
135
141
147
153
159

Eucalyptus lane-poolei Maiden **Salmonbark Wandoo** (SIVEA F.T.S.—)

A medium sized white gum found at the edge of the scarp and coastal plain near Perth in Western Australia.

Distribution

The species has a very restricted natural distribution which is mainly from near Perth to Pinjarra, is a distance of about 80km. It is found near the junction of the Darling Range and the coastal plain. There is also a small disjunct occurrence near Jurien Bay some 250km to the north. It is centred about latitude 32.25°S and at low altitudes of 30m of 120m.

Description

It is a small tree with a smooth pinkish white gum bark up to about 15m tall with a short crooked trunk branches above and has a spreading crown. It belongs to the group of common Western Australian species which have bisected cotyledons. The fruit is fairly large, 1cm or more in diameter, and has a strongly convex disc with short exserted valves.

Ecology

The species grows on gentle to moderate slopes with soils to low nutrient status, sandy texture and lateritic gravel. The climate is warm sub-humid with a mean maximum temperature of the hottest month of 30°C and mean minimum of the coldest month about 9°C. The mean annual rainfall is about 900mm with a pronounced winter maximum from May to August and very little rain at other times. At the Jurien Bay outlier the rainfall is only about 500mm.

The tree generally occurs in woodland or open forest fringing species such as *E. marginata, E. wandoo* or *E. calophylla.* In some stands it may be almost the only tree with an understorey of Western Australian Christmas tree, *Nuytsia floribunda.*

Conservation

Most of the original stands of *E. lane-poolei* have been cleared in agricultural and pastoral development. Those remaining are small and confined to a few localities near Serpentine and Badgingarra. There is considerable local interest in preserving some of the surviving patches.

Special Features

It is considered to have potential for amenity planting. The very limited and disjunct distribution suggest that is the pattern of a relict species.

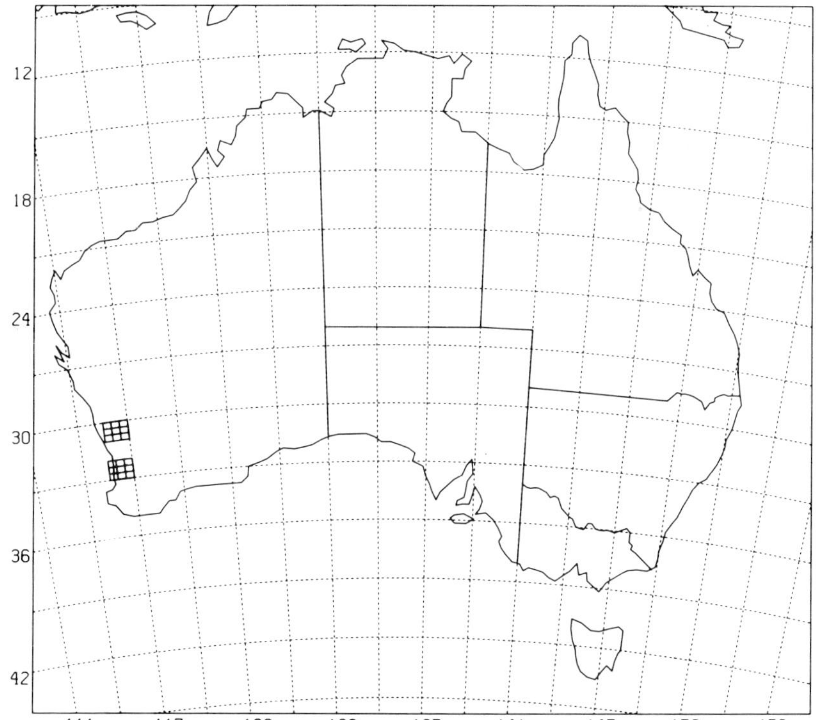
12
18
24
30
36
42
111
117
123
129
135
141
147
153
159

Eucalyptus largeana Blakely et de Beuzeville **Craven Grey Box** (SUDAA F.T.S. 191)

A large forest tree of northeastern New South Wales.

Distribution

Craven Grey Box has a scattered distribution in a belt about 250km long and 40-50km wide some 50-100km inland from the coast of NSW. The area is from the upper Hunter Valley in the south to the upper reaches of the Macleay and Manning Rivers in the north. Two well known occurrences are in the Avon State Forest near Craven and Copeland State Forest which is also in the Gloucester district. It occurs in latitudes of 31-33°S.

Description

E. largeana is a tall forest tree with a typical box bark and other characteristics of the box group of eucalypts. The adult leaves are rather dark green and the crown rather dense and the juvenile leaves are discolorous. The buds occur in clusters of 7 in both terminal and axillary groups. The tree sometimes exceeds 40m in height.

Ecology

It is a species found in hilly country at altitudes of 100-700m. The mean maximum temperature of the hottest month is about 27°C and the mean minimum for the coldest month is about 3°C and there are likely to be up to 20 frosts per year. The mean annual rainfall where the species is best developed is about 1100-1200mm with a summer maximum but with a fairly even distribution. It is commonly found growing on grey or brown podzolic soils of good depth which are often developed from shale. It is associated with species such as *E. microcorys, E. saligna, E. acmenioides, Syncarpia glomulifera* and Brushbox, *Tristania conferta*. It is a component of the vegetation community known as wet sclerophyll forest.

Conservation

Precise mapping of the species is difficult because it occurs in small areas and as a component of mixed tall forest. Most of the known occurrences are in state forests which are managed for hardwood production. It is not likely that it will be eliminated from these sites as it regenerates adequately following silvicultural activities.

Special Features

This species, with one or two others, is of particular interest because it belongs to the box group of eucalypts. This group is mostly found in woodland types of vegetation particularly on the inland and drier side of the Great Dividing Range. It seems possible that the two or three species like *E. largeana* have become adapted secondarily to the wet sclerophyll forest environment. The discolorous juvenile leaves are consistent with this idea. This is comparable with *Acacia bakeri* in the phyllodinous *Acacias* which occurs as a rainforest tree in Queensland, although most phyllodinous *Acacias* are found in very dry country. It could be argued, of course, that these are progenitors of the species of the more arid areas. The position remains unresolved.

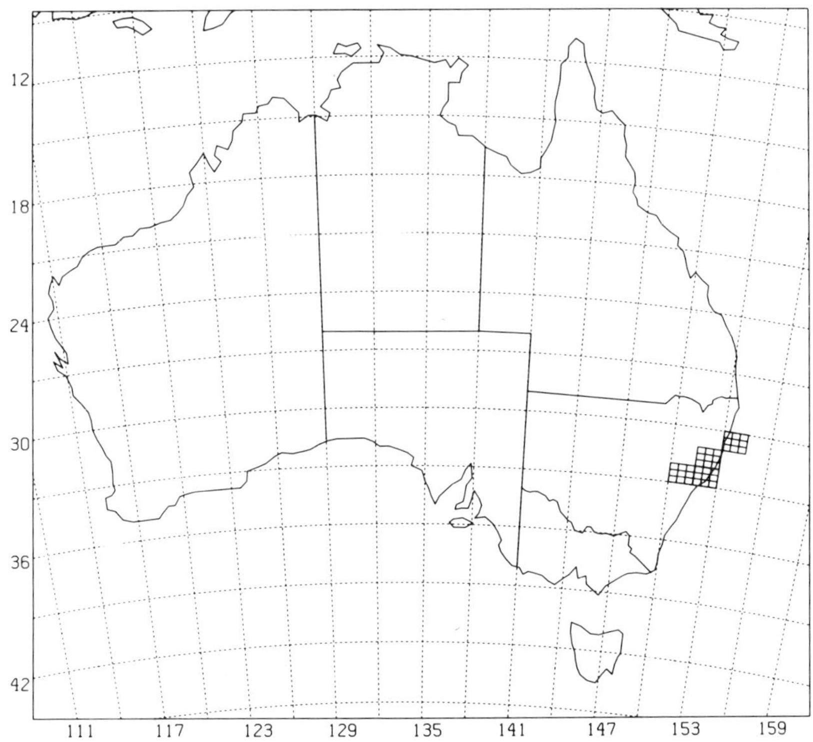
12
18
24
30
36
42
111
117
123
129
135
141
147
153
159

Eucalyptus luehmanniana F. Muell. **Yellow Top Mallee Ash** (MAKDB F.T.S. 32)

A moderate sized mallee ash of central eastern New South Wales.

Distribution

It is known from a number of locations in the vicinity of Sydney and south to Tomerong. it also occurrs in the Royal National Park east of Waterfall. The occurrences are not extensive.

Description

A small or moderate sized mallee with closely spaced stems of up to 8m tall. The foliage tufted to the ends of the stems. The adult leaves are rather thick and heavy and the buds have a distinctly pointed operculum. The buds are in clusters of more than 7 and the fruits are somewhat ribbed.

Ecology

Yellow Top Mallee Ash is confined to sites of low fertility through the latitudinal range 33-35°S. It is found from near sea level to about 300m altitude. The mean maximum temperature of the hottest month is about 27°C and of the mean minimum of the coolest month about 7°C. The annual rainfall is about 1100mm with a slight summer peak but generally well distributed. The soils on which it occurs are often skeletal, derived from Hawkesbury sandstone and of very low fertility. Massive rock outcrops are often present. Associated species are *E. obtusiflora, E. stricta* and *E. sieberi* together with many sclerophyll shrubs characteristic of the Hawkesbury sandstone area.

Conservation

It occurs on several locations in Kuringai Chase National Park as well as in the Royal National Park. Other occurrences are on crown land. Because of the nature of the sites it is not under significant threat and the occurrences in national parks should ensure its survival.

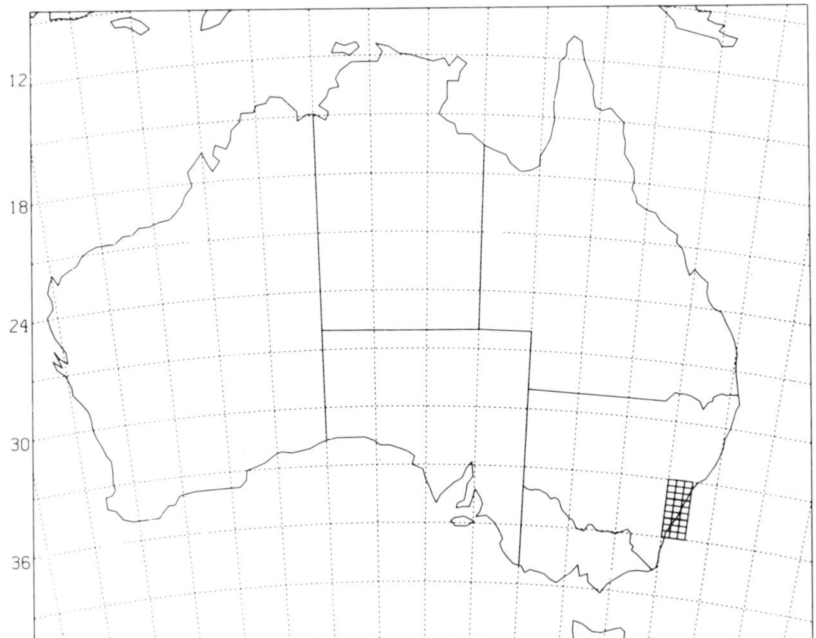
12
18
24
30
36
42
111
117
123
129
135
141
147
153
159

Eucalyptus megacornuta Gardner **Warted Yate** (SICBD F.T.S.—)

A small tree from the Ravensthorpe area of Western Australia.

Distribution

The species has been recorded from a few sites on the Ravensthorpe Range and areas nearby a little east of Ravensthorpe. A stand recently found was on a bank of a creek at the foot of the range and there are probably other sites which could be found by detailed study.

Description

It is a small tree up to about 7-8m in height with a smooth grey-brown bark. The buds and fruits are particularly striking with the very long operculum being coarse and markedly verrucose. The peduncle is very thick and flattened supporting three flowers. The fruit is large and campanulate. This species is very closely related to *E. burdettiana* which has similar buds and fruits but which is a smaller mallee-like species.

Ecology

As it is known from so few sites and these differ in detail, it is not possible to generalise about the species ecological requirements. It occurs around 100m at a latitude of abour 34°S. The mean maximum temperature of the hottest month is around 28°C and the mean minimum of the coolest month about 5°C. A few frosts occur each year. The rainfall is about 450mm with a prominent winter peak and a very dry summer. *E. megacornuta* occurs on sites of small extent in a few scattered localities in the Ravensthorpe Range where, in general the original vegetation remains. It is a member of low forest or dense woodland communities where the associated species are sclerophyll shrubs.

Conservation

Because the known occurrences are so limited will always be a risk that any single stand could be eliminated. However, under current land use patterns this is not a foreseen threat. It would be desirable to attempt to locate other stands of the species so that an adequate system of reserves can be set aside.

Special Features

E. megacornuta is of interest because of its rarity and because it is part of a super-species including *E. burdettiana*. it seems likely that a survey in the area between Ravensthorpe and Fitzgerald River would reveal more stands of the species or related populations. Its distribution suggests that it is a species that was spread widely and then contracted into small refuges.

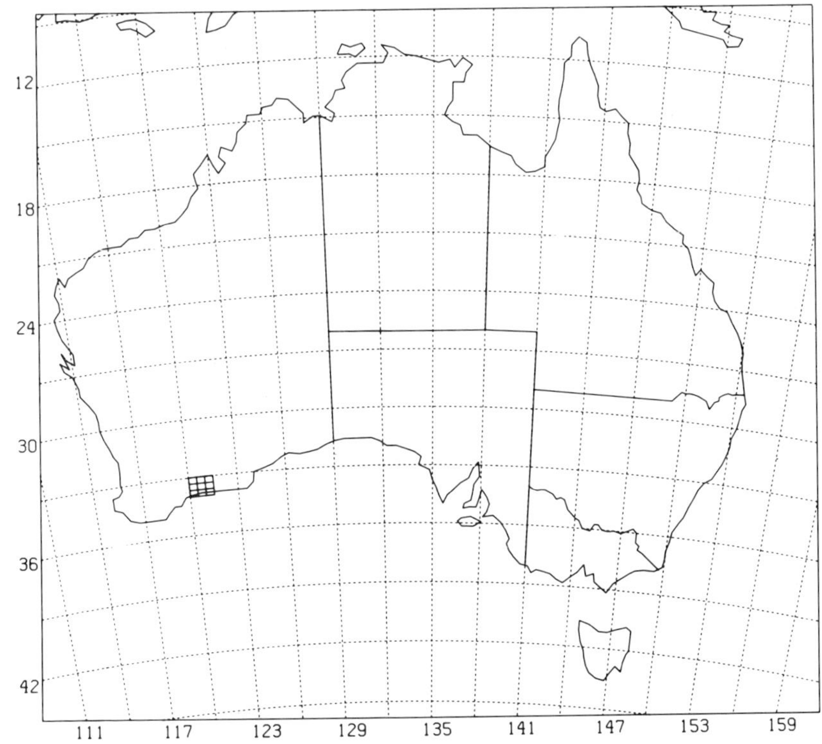
12
18
24
30
36
42
111
117
123
129
135
141
147
153
159

Eucalyptus michaeliana Blakely **Hillgrove Gum** (SNI:A F.T.S. 166)

A sizeable gum in the northeastern part of New South Wales.

Distribution

E. michaeliana has a much fragmented distribution with the separate sites having a relatively limited extent. The most extensive occurrence is in the Hillgrove old mining town area east of Armidale. it occurs in the vicinity of Wyong particularly near St Albans and there is also an occurrence in the ranges near Mt Ballon close to Queensland. The stands in the separate localities are of moderate size only.

Description

A tall gum up to 30m in height and with a trunk diameter of over 1 metre on the biggest trees. The bark is mottled greyish-white and the tree has a bushy crown. Of particular interest is the unusual form of the inflorescence in which the separate flower clusters occur in a raceme in leaf axils.

Ecology

Ecological conditions under which the species grows vary considerably across its range. It is found from latitude 28° to about 33.25°S and at altitudes close to sealevel in the Wyong area and over 1200 metres in the northern stands. The mean maximum temperature for the hottest month is 27°C and the mean minimum temperature for the coldest month about 5°C in the south of its range and 0°C in the north. The northern sites occur where there are moderate falls of snow in some years. The mean annual rainfall is about 1000mm with a summer maximum and a fairly even yearly distribution. It occurs in sclerophyll forest sometimes with other eucalypts such as *E. deanii* and *E. saligna* or on drier sites *E. globoidea* or *E. laevopinea*.

Conservation

In the Hillgrove area there has been very substantial clearing for pasture development and some of the best stands are now along road verges where upgrading of the road is in progress and likely to continue. In this area, where it occurs on flat ridges extending into the dissected valleys on the east of the tableland, there are probably other occurrences which would be revealed by detailed survey. A good deal of clearing has also taken place on the sites in the St Albans area. At present it is not under immediate threat of extinction but it is desirable that areas be included in appropraite reserves to ensure its long term survival.

Special Features

The unusual morphology of *E. michaeliana* makes it of interest. Also of particular interest is its very fragmented disjunct distribution on lowland and highland sites over a distance of some 800-1000km. It is a very good example of disjunct distribution of a highly distinctive species which has no very close relatives which very strongly indicates that there was a change in distribution following climatic fluctuations in the past.

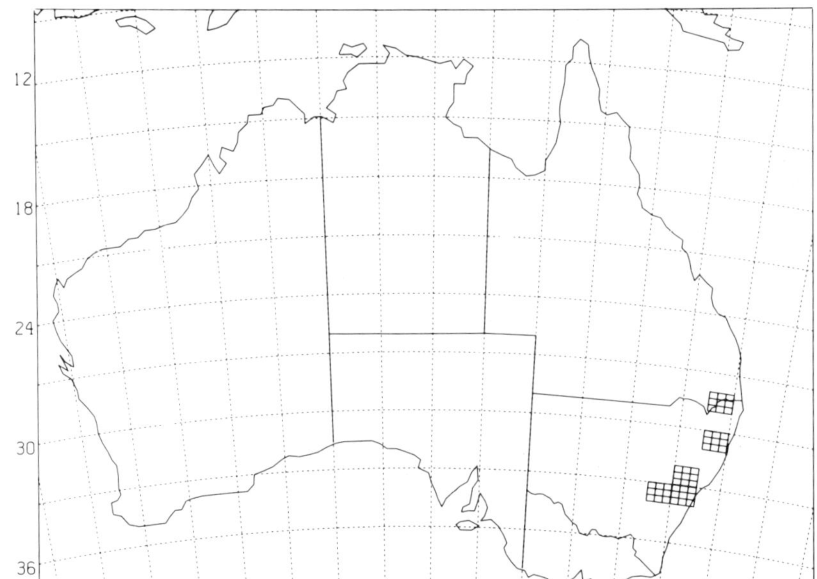
12
18
24
30
36
42
111
117
123
129
135
141
147
153
159

Eucalyptus mitchelliana Cambage **Mt Buffalo Gum** (MAKLA F.T.S. 90)

Mt Buffalo Gum is a moderate sized tree with attractive appearance and occurs in Victoria.

Distribution

E. mitchelliana is confined to two or three small localities on the top of Mt Buffalo east of Wangaratta in Victoria. It occurs both near the Chalet and on some nearby ridges, but the total distribution is less than one hundred hectares.

Description

It is a rather graceful tree with gum bark up to about 18m and a trunk 60-70cm in diameter. The leaves tend to be somewhat pendulous and form a dense crown. The buds are distinctive with a sharply conical operculum and occur in clusters of more than 7. The fruits are small and truncate-globular. It is a highly distinctive species, the nearest relatives of which are *E. kybeanensis* and *E. stellulata*.

Ecology

It grows in latitude 36.25°S and at an altitude of about 1300m on Mt Buffalo in Victoria. The mean maximum temperature of the hottest month is about 20°C and the mean minimum of the coldest month about 0°C. There are many frosts each year and snowfalls are heavy during winter and snow lies on the ground for several weeks at a time. The rainfall approaches 2000m a year with a winter maximum but reasonably well distributed throughout the year. Other eucalypts such as *E. dalrympleana* and *E. pauciflora* occur nearby and in the understorey various *Acacias* and other sclerophyll shrubs are found.

Conservation

It is entirely within the Mt Buffalo National Park where preservation should be assured. It appears to regenerate quite well amongst the granite rocks characteristic of the sites on which it occurs.

Special Features

It is a very distinctive species without very close relatives and with a highly restricted distribution. The fact that it is fairly readily recognised suggests that this may be the sole occurrence of the species. If that is the case, it is one of the extremely limited species which has a relict pattern which is so characteristic of many of the eucalypts. It grows quite vigorously in cultivation and is a useful ornmental tree.

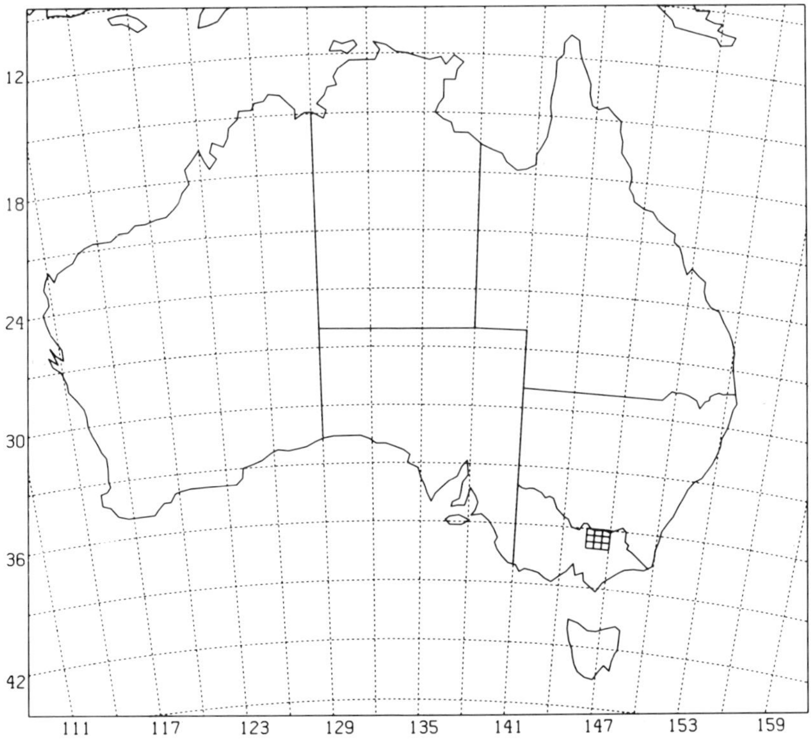
12
18
24
30
36
42
111
117
123
129
135
141
147
153
159

Eucalyptus morrisbyi R.G. Brett **Morrisby's Gum** (SPINK F.T.S. 52)

E. morrisbyi is a small to medium sized tree up to about 20m in height. It has a rather umbrageous crown and is found in the vicinity of Hobart in Tasmania.

Distribution

E. morrisbyi is now only found in two sites, one near Cremorne in Tasmania about 15km south east of Hobart and the other a very small remnant patch at Geilston Bay. In the Cremorne area there are a few trees along the roadsides and on the foreshores of Pipeclay Lagoon. All these stands are very small and one patch of about 2ha is in a paddock currently being grazed. There are a few roadside specimens.

Description

It is a smooth barked gum related quite closely to *E. gunnii.* It has typical orbicular juvenile leaves which are quite glaucous and buds and fruits in clusters of 3.

Ecology

It is found mainly near sealevel on gentle slopes with sandy soils derived from mudstone. The climate is sub-humid and very similar to that of Hobart. The mean maximum of the warmest month about 23°C and the mean minimum of the coolest month about 5°C. There are probably no frosts on the site near the sea but inland there may be five a year. The rainfall is about 600mm a year and very light snow may fall occasionally. It is found in pure stands with *E. viminalis, E. ovata, E. amygdalina* and *E. tenuiramis* nearby and occasionally with some *E. globulus.*

Conservation

The species had a very limited distribution. Most of the stands have been removed and the existing occurrences are limited to farmland or roadside verges. Some regeneration is seen on the roadside sites but there is virtually none in the grazed paddocks.

Changes in land use could very easily eliminate the species.

Special Features

It is obviously closely related to *E. gunnii* and has some affinity with other Tasmanian species such as *E. urnigera* and *E. subcrenulata* as well as the mainland species *E. glaucescens.* It is tempting to suggest that *E. morrisbyi* has developed as a distinct species as the surviving coastal part of a population of eucalypts from which *E. gunnii* has been derived. *E. gunnii* is characteristic of sites of higher altitude than *E. morrisbyi* and occurs in areas of higher rainfall in central and northwestern Tasmania. *E. morrisbyi* occurs in a part of Tasmania where access is good and it is not likely that many more remnant stands of the species will be found. There have been quite extensive searches for the species. This species is very close to extinction and could easily be lost completely in the natural state unless some steps are taken to set up appropriate reserves. The Geilston Bay site is reserved but it has only a few mature trees and this is scarcely adequate to ensure the survival of the species.

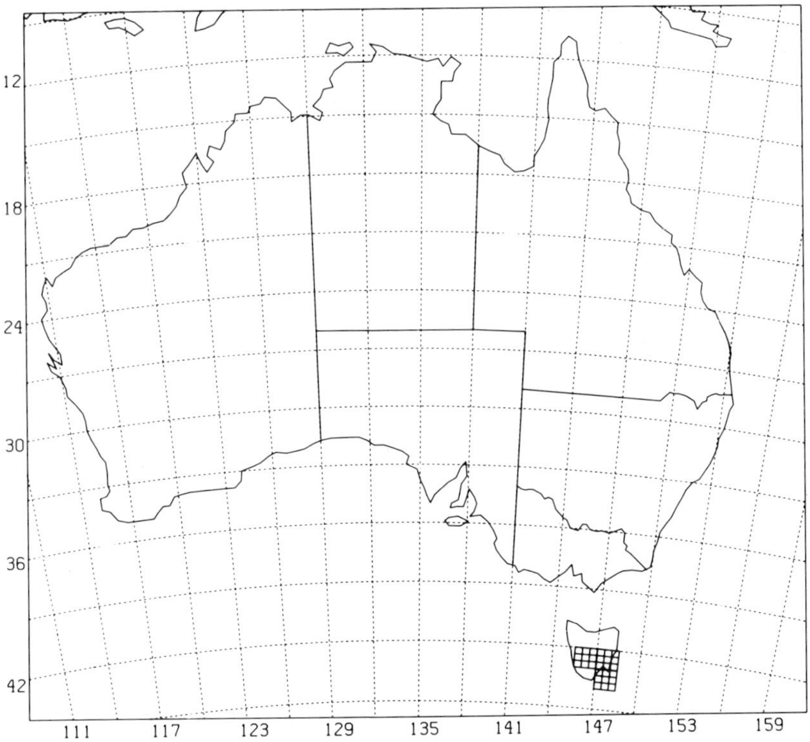
12
18
24
30
36
42
111
117
123
129
135
141
147
153
159

Eucalyptus neglecta Maiden **Omeo Gum** (SPIAA F.T.S. 88)

This is a small tree of damp sites in eastern Victoria.

Distribution

It occurs in small patches in an area from Omeo to Bright in Victoria and southwards to Jamieson. It is mostly found at high altitudes but near Bright descends to around 300m.

Description

A small tree up to about 7 or 8m in height with a rather bushy habit and branches well down on the stem. The juvenile leaves are quite broad sessile and orbicular in shape. The buds and fruits are sessile and in clusters generally of about 7 or more. The bracts covering the young flower cluster in bud are very broadly keeled and the keel in this species may be the largest in the genus.

Ecology

This species mostly occurs at high altitudes on rather swampy sites or moist banks of rivers or small streams. It is found at about latitude 37°S and at altitudes from about 500-1200m. The mean maximum temperature of the warmest months is around 23°C and the mean mimimum of the coolest month about 1°C. Rainfall is about 1200mm a year. The sites are often restricted and the species occurs in narrow zones along streams with typical wet sclerophyll species nearby. *E. ovata* and *E. viminalis* are usually associated with *E. neglecta* as are *Acacia melanoxylon* and tree ferns. At times Omeo Gum forms an understorey to the taller eucalypts.

Conservation

Most of the sites are in relatively inaccessible country and some occur on crown land. The best stands are probably on private property and therefore long term conservation will depend on some form of reservation. There are probably more stands than have been recorded since the area in which it occurs is rather difficult to survey in detail.

Special Features

Like a number of other restricted species of southeastern Australia, it appears to be a species capable of only moderately vigorous growth. Its fragmentary occurrence and its association with moist damp sites suggests that it could have been much more widespread in a cool pluvial period in the past. It has proved to be very cold-resistant to the special climatic conditions that occur in the southeast of the United States. In this area during a period of normally mild temperatures sudden freezing conditions can occur for 2 or 3 days and this kills most eucalypts. It shares this feature to some extent with *E. parvifolia* and the two species are more or less unique in this regard.

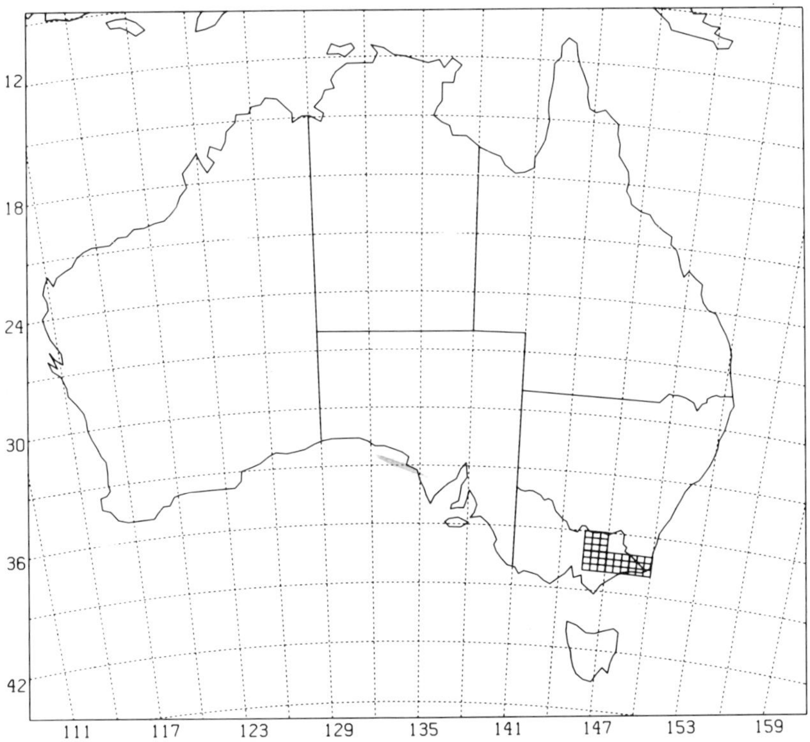
12
18
24
30
36
42
111
117
123
129
135
141
147
153
159

Eucalyptus olsenii Johnson & Blaxell **Olsens Gum** (MAJ:O F.T.S.—)

This is a small tree with a straight stem found in mountain country south of Braidwood, New South Wales.

Distribution

It is known from only 2 or 3 sites in the dissected mountainous area southeast of Braidwood in the vicinity of Dampier Trigonometric Station.

Description

It is a small tree up to 10m in height with smooth bark on the trunk and branches and a small stocking of rough bark at the base. The juvenile leaves are thickly covered with radiating hairs. The fruits are rather large, globular to barrel-shaped and somewhat wrinkled when dry. It has features of both stringybark and ash-type eucalypts.

Ecology

The species occurs on rather stony sites on one or two small ridges in the vicinity of Dampier Trigonometric Station. The mean maximum temperature of the hottest month is about 27°C and the mean mimimum temperature of the coldest month about 2°C. Rainfall is about 1200mm a year. Other species that grow nearby are *E. dalrympleana*, *E. sieberi* and a number of other species, together with the common sclerophyll shrubs of this type of forest.

Conservation

It is in very inaccessible areas and there may well be other stands yet to be discovered although survey in this area is difficult. it occurs within the Deua National Park and this, together with the general inaccessbility of the sites should ensure its preservation.

Special Features

The species is of particular interest because it has no close affinities. It is clearly related to stringybarks on the one hand and ashes on the other, but it is not in any sense intermediate between them. The occurrence in a very small area of a species with such unusual and highly distinctive features is of great interest and supports the view that it is a relict occurrence. It is regenerating satisfactorily in the natural habitat and can be raised readily in cultivation.

12
18
24
30
36
42
111
117
123
129
135
141
147
153
159

Eucalyptus paliformis Johnson & Blaxell **Wadbilliga Ash** (MAKIM F.T.S. 195)

This is a slender, single-stemmed tree up to about 15m in height.

Distribution

It occurs in 2 or 3 small stands southeast of Wadbilliga Trignometric Station in rough dissected areas southeast of Cooma in the upper Tuross River area.

Description

A slender ash seldom much more than a large sapling in size. The juvenile leaves are dark green rather than glaucous and the adult leaves are shiny. The buds are in clusters of 7 and the fruits contracted at the orifice and more or less truncate globular. It has affinity with *E. fraxinoides*. The shiny black seed of *E. paliformis* with its reticulated sculptured surface is different from *E. fraxinoides*. It is also without lignotubers which is characteristic of many species of ash group.

Ecology

It is known only on a ridgetop in an area of a few hectares in two or three patches although it probably occurs on some other ridges nearby which await exploration. It is in latitude 36°S and at an altitude of about 1000m. The mean maximum temperature of the hottest month is about 22°C and the mean minimum temperature of the coldest month is -2°C. Frosts are frequent each year and heavy snowfalls often occur. The rainfall is around 900mm a year. It is found on shallow sandy soils over fine-grained sandstones. Species nearby are *E.fraxinoides*, *E. nitens*, *E. kybeanensis* and *E. pauciflora*, together with a substantial number of sclerophyll shrubs. Regeneration occurs effectively in the area.

Conservation

Apart from activities such as construction of fire trails, it is unlikely that the species because of the nature of the site would be disturbed for any foreseeable land use. it is in a quite inaccessible area which is within the Wadbilliga National Park. Its preservation should be assured as a result of this.

Special Features

Like a number of other species it is of particular interest because of its highly restricted known distribution. It is possible that a few more stands may be found. It is quite a handsome small tree. While it is somewhat like *E. fraxinoides* it is a distinct species.

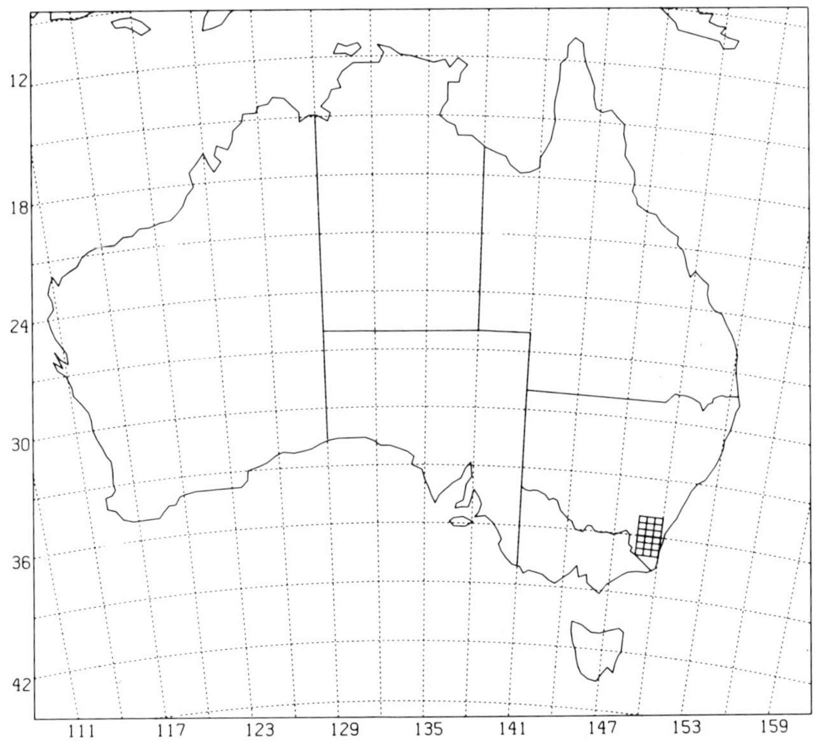
12
18
24
30
36
42
111
117
123
129
135
141
147
153
159

Eucalyptus parvifolia Cambage **Small Leaved Gum** (SPIBA F.T.S. 91)

A small roundheaded tree growing on wet or even somewhat swampy sites in shallow valleys near the edge of the tableland in the southeast of NSW.

Distribution

It occurs in small stands in valley bottoms in a narrow zone from near Badja and Kybean to southeast of Nimmitabel. The natural stands were never extensive or widely distributed. Many stands have been cleared for agriculture and the surviving stands are mostly on private land with a few trees remaining on roadside verges.

Description

A small, often crooked, round-headed tree 8-12m in height with a smooth greenish or grey bark. On older and larger trees there is a stocking on the butt of hard, furrowed, dark and somewhat tessellated bark.

The adult leaves are shortly stalked or nearly sessile. The leaves are small, often not more than 5cm long x 1cm wide. The buds and fruits are small and in tight clusters of 7. It is a distinct species with no close relatives. In the field it resembles superficially, *E. stellulata* which is in a different subgenus.

Ecology

E. parvifolia occurs in latitude 36°S at an altitude of about 1100m in shallow valleys. These valleys which are near the scarp of the tableland are frost hollows and fogs are common. The soils are mildly acid, brown meadow type and often somewhat swampy. The climate is sub-humid with cool summers and cold winters. There are up to 100 frosts a year and always some snow which may lie on the ground for several days. The mean annual rainfall is about 750mm. it is sometimes in small pure stands but is often mixed with *E. stellulata* and *E. pauciflora* and with a wet tussock grass of *Poa caespitosa* or heathy undergrowth.

Conservation

The survival of this species depends on a change in land use patterns. Further development for agriculture could eliminate the species entirely except for a few roadside specimens which would be vulnerable to road upgrading.

It regenerates well only if grazing is withheld for a time. Most of the existing stands are probably already recorded and little additional survey is necessary.

The international interest in the species places it at some risk because of the economic value of seed. Seed collection if done by branch cutting can be destructive.

Special Features

E parvifolia has proved to be one of the most cold resistant of the eucalypts when planted in the northern hemisphere, especially in the UK and parts of France. Its preservation as a gene resource already has demonstrable value.

The physiological feature of the species in resisting low temperatures is anomalous since it does not occupy the coldest Australian sites. Areas with frost hollows in the Kosciusko region and in Tasmania are certainly colder than the natural habitat of *E. parvifolia*. There is a good reason to suppose this is a relict species from a cooler, more pluvial period.

Interspecific hybrids with either *E. rubida* or *E. viminalis* are common at the periphery of the stands and this fact must be taken into account in a reproduction programme.

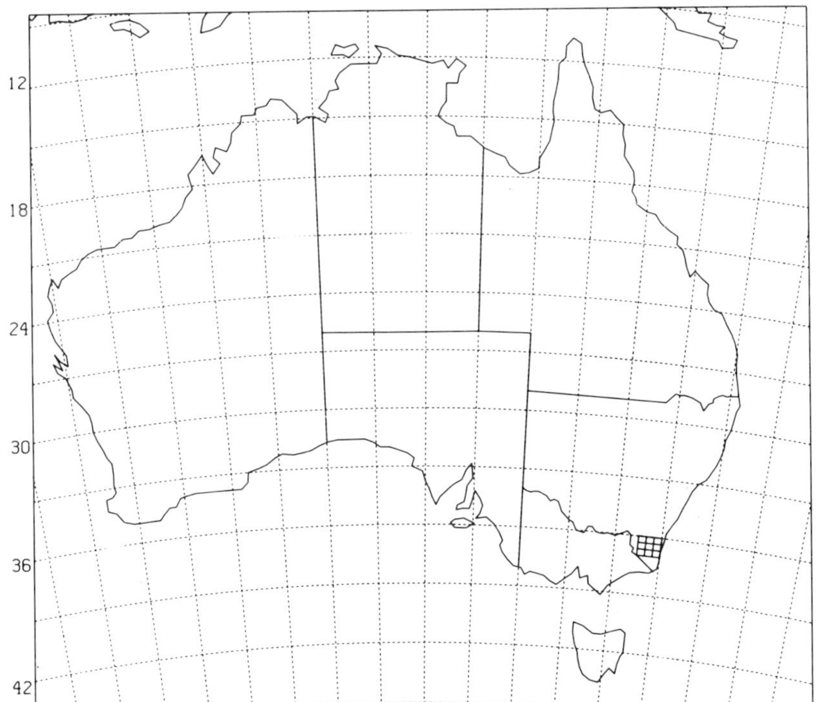
12
18
24
30
36
42
111
117
123
129
135
141
147
153
159

Eucalyptus pendens Brooker **Badjingarra Mallee** (MADAD F.T.S. 183)

This is a very small, open-crowned, pendulous, thin-stemmed mallee with a distinctive appearance. It has been recognised recently on sites which have become accessbile with new highway development north of Perth towards Geraldton in Western Australia.

Distribution

The species was described from material collected in 1969 by M.I.H. Brooker and its distribution is not yet well known. It has been recorded in an area about 200km north of Perth between Gin Gin and Badjingarra. There are very few recorded sites but the area has not been adequately surveyed. The few known occurrences are small groups of a few trees.

Description

It is a thin-stemmed mallee up to about 5m in height with very pendulous drooping branches. It has a very open thin crown with only a few leaves.

Ecology

It occurs on undulating land of very low fertility some 30-40km from the west coast of Australia at a low altitude. It occurs in heathy vegetation with genera such as *Xanthorrhoea*, *Hakea* and *Banksia*. It is prominent in this community because it stands out above the surrounding vegetation.

The summers are hot and the winters cool with a yearly rainfall of about 500mm with a marked winter peak.

Conservation

The known occurrences are on roadside verges, which are subject to risk from realignment of roads, and on farmed land where clearing is in progress. It is quite possible however, that there are other stands yet to be detected. Reservation to ensure the survival of this species is important but this should be preceded by survey to establish the full extent of the species.

Special Features

Along with *E. sepulcralis* and a weeping form of *E. pauciflora* at Adaminaby, NSW, it is one of the most pendulous and open-crowned species of the genus. This unusual form is of particular interest and it is a characteristic feature of species in the area.

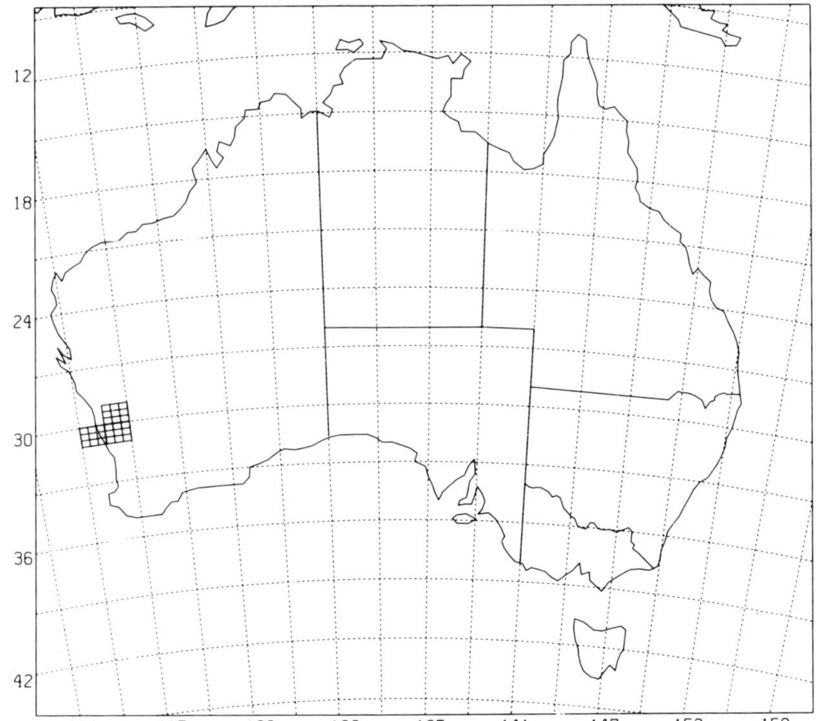
12
18
24
30
36
42
111
117
123
129
135
141
147
153
159

Eucalyptus pulverulenta Sims **Silverleaved Mountain Gum** (SPINQ F.T.S. 28)

This eucalypt is a small straggling tree or tall shrub found in southern New South Wales east of Cooma, and also west of Sydney.

Distribution

It is known from a few localities in two general places, firstly near the Lithgow-Jenolan Caves road west of Sydney and secondly in the vicinity of Cowra Creek east of Bredbo, which are both in New South Wales. So far as is known there are no intermediate stands between these two quite disjunct occurrences. There are several stands in the general area of the southern location. Some of those recorded in the northern area can no longer be found in the field.

Description

This is a species which retains juvenile foliage throughout its life. The leaves are round, sessile and highly glaucous. Its buds and fruits are in clusters of 3. The stems are rather crooked and branch at odd angles and the whole tree is seldom more than 5 or 6m in height.

Ecology

This species occurs on rocky sites in hilly country at altitudes of about 900m. The two occurrences are at latitudes 33°S and 36°S. The mean maximum temperature of the warmest month is about 27°C and the mean minimum temperature of the coldest month is about 0°C. There are a substantial number of frosts each year. The mean annual rainfall is about 800mm with a moderate summer peak. *E. pulverulenta* occurs on soils derived from siliceous metamorphoric rock. The soils are often skeletal and rock outcrops are common. While this species does occur in pure stands, at times it is found as an understorey to taller eucalypts such as *E. dives* and *E. mannifera*.

Conservation

It occurs in areas which are marginal for grazing or otherwise so rocky they are not likely to be disturbed. Some stands could be under threat by mining activities. The best existing stands are on what is probably privately owned land and conservation of these areas would require special arrangements with the owners. Some of the Cowra Creek stands are in an area which is being considered for reservation as a national Park.

Special Features

E. pulverulenta, like *E. kruseana,* is a very unusual species in both form and the way in which it preserves the juvenile type leaf throughout most of its life. It is valued as an ornamental species particularly in the United States where its foliage is very much in demand by florists. It is particularly suitable as an experimental plant for the production of interspecific hybrids and it has been subject to a large amount of experimental breeding research.

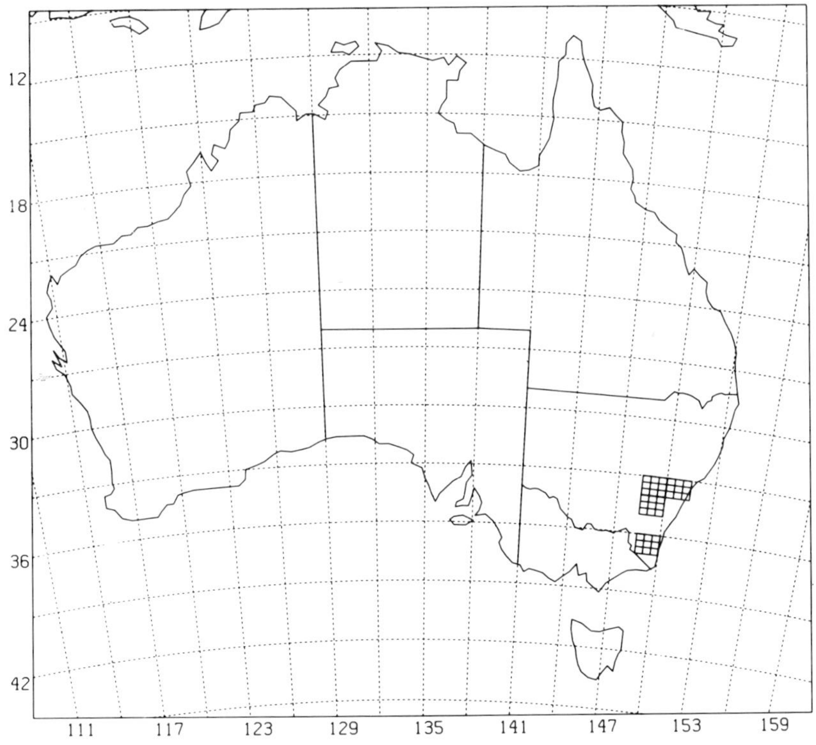
12
18
24
30
36
42
111
117
123
129
135
141
147
153
159

Eucalyptus pumila R.T. Baker **Pokolbin Mallee** (SNECH F.T.S. 210)

This eucalypt is known from only one stand, of less than a hectare, in the vicinity of Pokolbin in the Hunter Valley of New South Wales.

Distribution

The only record is a single stand in the vicinity of Pokolbin on the northern slopes of Brokenback Range in a state forest. It seems likely that there are two or three other small occurrences nearby in rugged areas with difficult access.

Description

E. pumila is a small mallee with slender stems reaching up to about 6m in height. The juvenile leaves are alternate and petiolate. The flowers are in clusters of 7 with the buds having a rather long operculum. The fruits have exserted valves and are a moderate size.

Ecology

This species occurs in latitude about 33°S and at an altitude of around 250m. The mean maximum temperature of the warmest month is about 31°C and the mean minimum temperature of the coolest month about 4°C. The yearly rainfall is about 750mm with a relatively even distribution. *E. pumila* occurs on steep slopes with sandy soil derived from sandstone and conglomerate both of which outcrop through the site. It occurs in a more or less pure stand with nearby trees of *E. maculata, E. fibrosa* sub-species *nubila, Callitris endlicheri* and *Casuarina,* together with various sclerophyll shrubs including *Acacia.*

Conservation

Because the species is of such a highly restricted distribution and at present is known from only one stand, it must be considered at some risk even if only through unlikely or unforseen events. It occurs in state forest and is not likely to be eliminated through any land use operations. Since it sprouts well from the root stock it persists well following the frequent fires.

Special Features

The species has attracted a lot of botanical interest because it has characteristics which are generally intermediate between those of the grey gums and the red gums. It is regarded as being an unusual species without close affinities although it has some similarities with both the grey and red gums. Its restricted occurrence and the distinctness of the species make it considerable botanical interest.

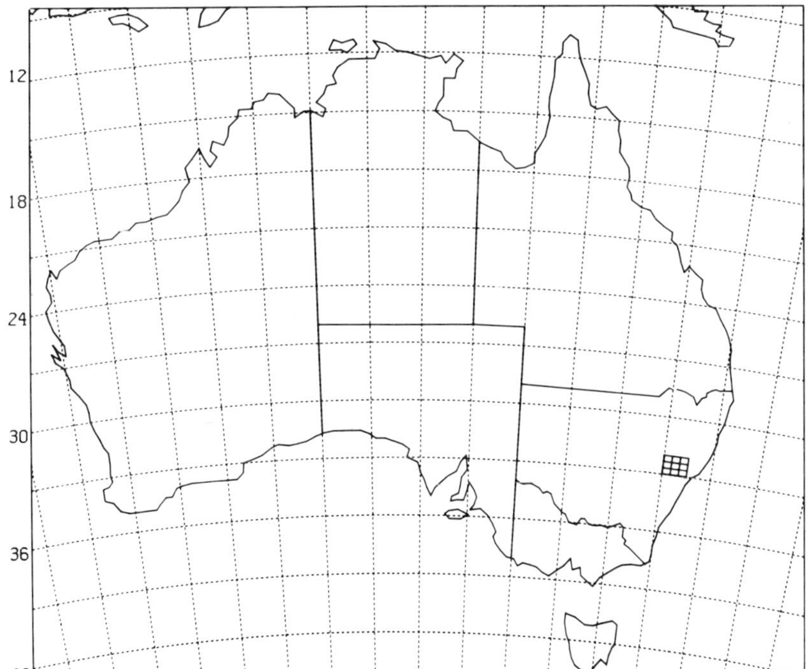
12
18
24
30
36
42
111
117
123
129
135
141
147
153
159

Eucalyptus remota Blakely **Kangaroo Island Mallee Ash** (MAKEB F.T.S. 128)

This is a small mallee ash from Kangaroo Island.

Distribution

It is confined to the western end of the high ridge of the island at a number of localities in the vicinity of Mt Taylor and Flinders Chase.

Description

It is a small multi-stemmed mallee up to about 4 or 5m tall with a rather dense crown carried terminally on the branches. The bark is persistent and roughly subfibrous but towards the ends of the branches it is smooth and decorticating. The leaves are rather coriaceous and dark green. The inflorescence has numerous flowers, frequently as many as 20, and the fruit is truncate globoid with a flat disc. It is quite closely related to *E. consideniana* of eastern Australia.

Ecology

E. remota occurs in about latitude 36°S and at altitudes of about 200m. The mean maximum temperature of the warmest month is about 25°C and the mean minimum of the coolest month about 10°C, with scarcely any frosts. Rainfall is about 600mm per year with a pronounced winter maximum. The sites are gently undulating and the soils are mainly heavy laterites with pale-coloured sandy podzols on lower slopes. This species forms a sclerophyll mallee scrub and is often associated with other mallees or stunted trees such as *E. diversifolia* and *E. obliqua*. The vegetation communities include sclerophyll shrubs such as *Banksia*, *Casuarina*, *Hakea* and *Xanthorrhorea*.

Conservation

Some of the stands have been cleared for agriculture but there are considerable numbers of stands within the boundary of Flinders Chase. The species survival should be assured within the Chase. Fires occur from time to time, but the species recovers well from burning because of capacity to shoot from the rootstock.

Special Features

There is particular interest in this species from the boigeographical viewpoint because of its relationship with the fairly widespread *E. consideniana* of eastern Victoria and southeastern New South Wales. These patterns suggest that *E. remota* is an outlier of the forerunners of *E. remota* and *E. consideniana*. The forerunner was left isolated in the favourable mild climate of Kangaroo Island and has been isolated sufficiently long to develop as a distinctive taxon. *E. remota* is now justifiably regarded as a separate species in much the same way as *E. argophloia* is separate from *E. bosistoana*.

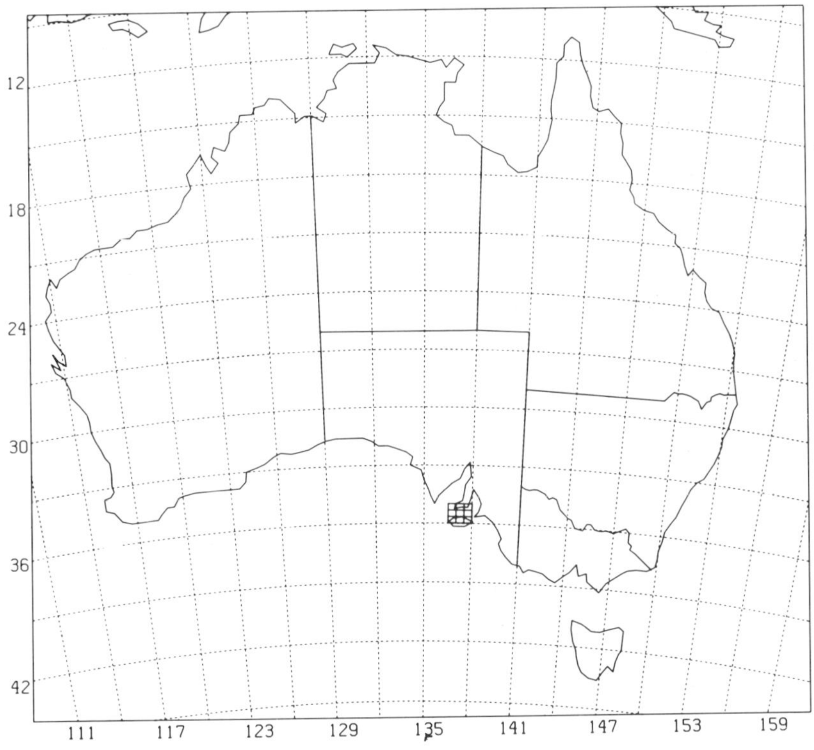
12
18
24
30
36
42
111
117
123
129
135
141
147
153
159

Eucalyptus rhodantha Blakely & Steedman **Rose Gum** (SIVEF F.T.S.—)

This is a small, low spreading mallee of Western Australia which has striking value as an ornamental plant.

Distribution

Rose Gum occurs in a narrow belt from Three Springs to Bolgart and south to Watheroo. The distribution is known only from remnant stands since, in most of this area, the original vegetation has been cleared. The only area with more than a few scattered trees is of about 3ha near Watheroo which is on uncleared grazing land.

Description

This is a low spreading mallee which grows to no more than 3m in height. At maturity there are several stems with smooth bark forming a rather dense crown. The leaves are opposite, sessile, rather broad, highly glaucous and blue-grey in colour. The buds are often solitary but are also in 3's. They are glaucous and quite large being up to 3-4cm in diameter. These buds are only a little smaller than the buds and fruits of *E. macrocarpa* which has the largest in the genus. The flowers are up to 6cm in diameter and have bright light-red stamens and prominent yellow anthers. The fruit is correspondingly large and is up to 4cm in diameter. It is a most attractive plant with flowers of considerable ornamental value. It is evidently related to *E. macrocarpa* but is distinct from that species in a number of ways.

Ecology

This species occurs as a component of heathy scrub and is found in small stands on grey sandy soil. It is associated with typically Australian genera as *Banksia, Casuarina* and *Hakea* as well as a wide array of other shrubby heathland species. *E. accedens, E. eudesmoides* and *E. tetrodonta* occur in the general vicinity. The summers are rather hot with a mean maximum temperature for the hottest month of 31°C and the winters cool with only occasional frosts. The rainfall is about 400mm per annum with a pronounced winter peak.

Conservation

The species is in a very precarious position. There are a few individuals on roadside verges which could easily be eliminated by realignment or upgrading of the roads. The only other stands are in an area of farmland which is used for grazing and there is no eucalypt regeneration. It is unlikely that more stands will be found since it is in an area which is subject to extensive wheat farming and grazing.

Special Features

E. rhodantha has been the subject of debate about its systematic status. There is a large amount of variation in the specimens assigned to this species and some of them appear to be intermediate with *E. pyriformis.* Whether this is due to interspecific hybridising or to other variation is not clear because of the lack of sufficient living speciments to examine. The Watheroo site appears to be the most distinctive form of the species and does not possess the variation which suggests affinity with *E. pyriformis.* Pryor and Johnston, when considering this species, were not aware of this distinct stand and believed from the specimens available that it may have been an interspecific hybrid. This does not seem to be the case but rather that it is a highly restricted species which is distinct from other eucalypt species although it has some relationship with *E. macrocarpa* and *E. pyriformis.* Rose Gum has very considerable ornamental value.

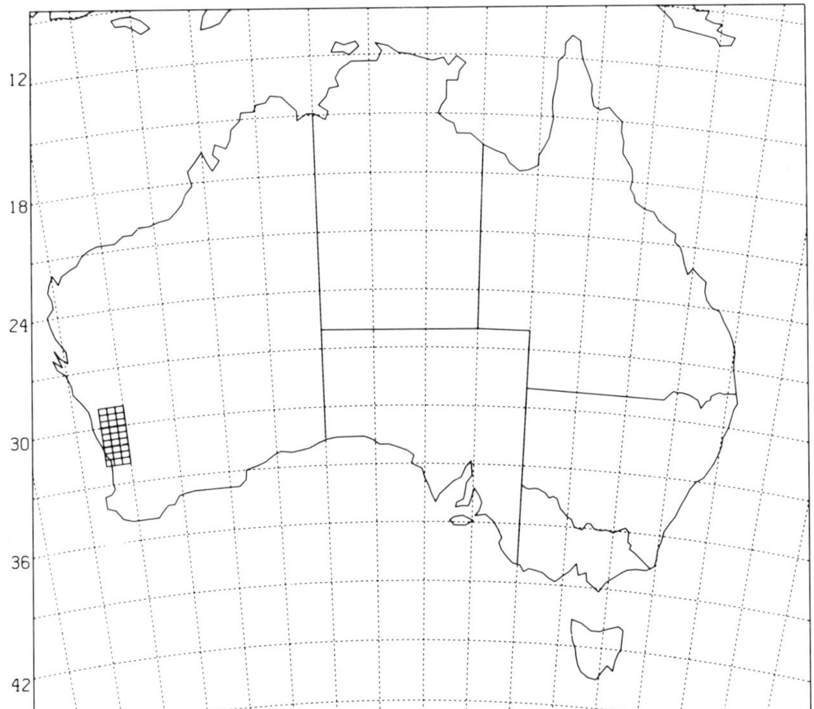
12
18
24
30
36
42
111
117
123
129
135
141
147
153
159

Eucalyptus risdonii Hook.f. **Risdon Peppermint** (MATEB F.T.S. 151)

It is a medium sized tree found in the vicinity of Hobart.

Distribution

All records for the occurrence of this species are close to Hobart, particularly on the Risdon or northern side of the Derwent River.

Description

This species is smooth-barked. It is a species in which the juvenile leaves persist for a very long time and the adult leaves relatively seldom develop in the natural environment. The juvenile leaves are highly glaucous, sessile, opposite and perfoliate. The buds and fruits are in clusters of 7 or more and the fruits are hemispherical with a flat rim. On poorer sites it becomes a mere mallee of 2-3m in height but it can form a tree of up to 17-18m in height.

Ecology

Risdon Peppermint occurs in about latitude 43°S at altitudes from sealevel to about 200m. The mean maximum temperature of the hottest month is about 22°C and the mean minimum temperature for the coolest month about 4°C. A few frosts occur on the sites away from the sea and rainfall is about 600mm a year with a moderate winter peak. Very light snow may occasionally occur but the sites are too close to the sea for snow to remain on the ground. It occurs on moderate to steep slopes and on well drained skeletal soils or at times medium quality podzols derived from mudstone. Associated species are *E. viminalis*, *E. globulus* and *E. amygdalina* together with sclerophyll shrubs characteristic of this type of eucalypt forest.

Conservation

There are a number of stands of the species in the general area and, although some of it has been cleared, most of it is on sites too poor for agriculture. Many of the surviving stands are on private land. Perhaps the greatest threat to the species is urban development and it is very desirable that adequate stands be reserved. Some small occurrences are already in small reserves but it appears that the species needs further reservation to make its position more secure.

Special Features

E. risdonii is unusual in the Monocalyptus sub-genus in having long-persistent juvenile foliage. Where this does occur in eucalypts it is mostly in other sub-genera. Because of its glaucous appearance and opposite leaves, it has developed a place as an ornamental plant. It is raised from seed and can be cultivated easily.

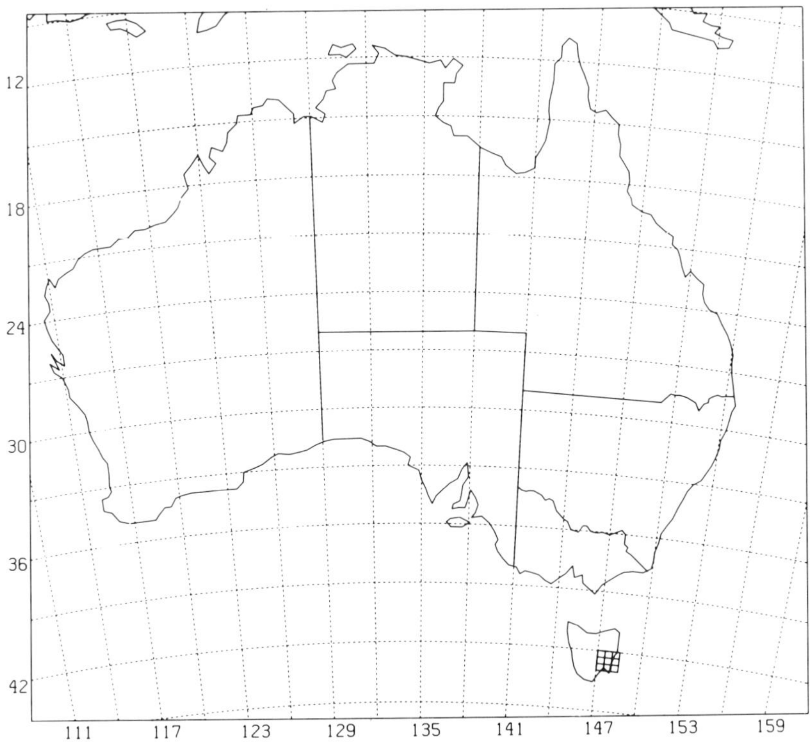
12
18
24
30
36
42
111
117
123
129
135
141
147
153
159

Eucalyptus rummeryi Maiden **Steel Box** (SUAAA F.T.S. 168)

This is quite a tall tree from northeastern New South Wales.

Distribution

Steel Box occurs in small scattered stands on the eastern slopes from the northern tablelands in the vicinity of Casino, south to east of Dorrigo in a zone about 200km long and 30 to 40km wide. Through this zone however the stands are quite limited and scattered.

Description

It is a moderate to large sized tree with a straight trunk and reaches a height of some 40m. The bark is typically fibrous box bark and is found on most of the trunk and large branches. The juvenile leaves are discolorous, and the adult leaves are a typical box type. The flower clusters are in a paniculate arrangement and the fruits quite small and have sunken valves.

Ecology

The tree occurs in wet sclerophyll forest and often at rainforest margins. The sites are in about latitude 29-30°S and altitudes of from 100 to about 400m. The mean maximum temperature of the hottest month is about 29°C and the mean minimum temperature of the coldest month about 5°C. There may be a few frosts. This species occurs on slopes and tops of broad ridges in hilly and undulating country. The soils are moderately fertile and are often derived from basalts, although other substrate is involved. The soils are often clayey with an amount of forest humus. There are often Hoop Pine, *Araucaria cunninghamii* and Crows Ash, *Flindersia australis* nearby and the eucalypts in the vicinity include *E. amplifolia, E. maculata, E. moluccana,* and *E. paniculata.* An understorey of rainforest species is frequently common including members of the families *Sapindaceae* and *Euphorbiaceae.*

Conservation

Most occurrences of the tree exist in state forests being managed for continuous production from the native species. It regenerates adequately and, under the existing system of management, its persistence is ensured.

Special Features

E. largeana and *E. rummeryi* are the only box species which do not occur in relatively dry environments. *E. rummeryi* occurs in marginal rainforest sites suggesting secondary adaptation to the wetter sites from an ancestory adapted to drier sites. This view is, however, only speculative.

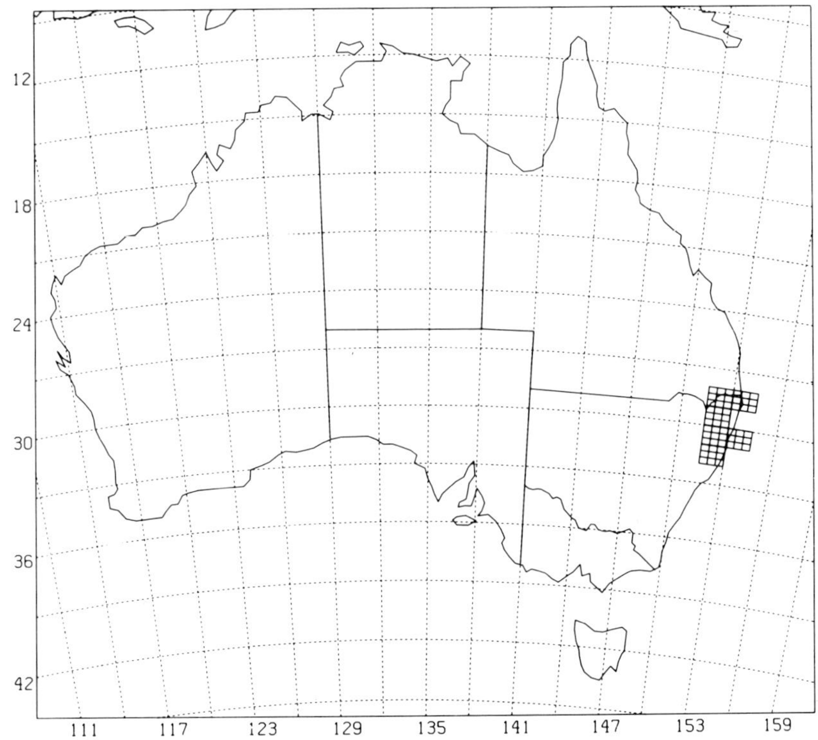
12
18
24
30
36
42
111
117
123
129
135
141
147
153
159

Eucalyptus saxatilis Kirkpatrick **Mt Wheeler Mallee** (SPINJ F.T.S.—)

A small, shrubby tree or mallee of eastern Victoria.

Distribution

This species occurs in mountainous areas on the western bank of the lower Snowy River on hilltops at the edge of rocky gorges and steep ridges. There are three locations in the area but detailed survey could reveal more since access is difficult in most of the area.

Description

This eucalypt is a tall shrub or small tree often of mallee form reaching about 8m in height. It has a smooth bark, somewhat glaucous adult foliage, very glaucous juvenile leaves and a somewhat bell-shaped fruit. It shows affinity with *E. glaucescens*. The most distinctive form of *E. saxatilis* is at the Mt Wheeler site and these specimens are quite distinct from *E. glaucescens*. The buds and fruits are in clusters of 3.

Ecology

It occurs at altitudes of about 900m on very rocky sites which have massive porphyry outcrops and skeletal soils. It is often associated with a depauperate form of *E. smithii* and occurs in latitude 37°. The mean maximum temperature of the hottest month is about 28°C and the mean minimum of the coldest month about 2°C. Mean annual rainfall is about 700mm with a moderate winter peak.

Conservation

The sites on which it occurs are unlikely to be disturbed unless by some unusual engineering need. The Mt Wheeler occurrence is within a flora reserve and the species is under little immediate threat.

Special Features

E. saxatilis is variable in its form and has obvious affinity with *E. glaucescens* and, perhaps in the more extreme forms, to *E. maidenii*. It is a species with a highly restricted occurrence. In some of the stands it may be genetically blurred by some infiltration from *E. glaucescens*. More extensive study, especially if other stands are found, might add to the understanding of the biological background of this species.

12
18
24
30
36
42
111
117
123
129
135
141
147
153
159

Eucalyptus scoparia Maiden **Wallangarra White Gum** (SPECM F.T.S. 141)

A small white gum which is found on the Queensland-New South Wales border.

Distribution

E. scoparia has a very restricted distribution being found just east of Wallangarra in Queensland where it is known from two large granite outcrops, Mt Norman and Castle Rock. There are rather similar sites, just over the border in NSW, on which the species may occur.

Description

A slender white gum up to about 20m tall with drooping branches and rather long narrow leaves. The juvenile foliage is green and sessile and the buds and fruits are in clusters of 7.

Ecology

Wallangarra White Gum occurs on skeletal soils and fissures on the granite outcrops of Mt Norman and Castle Rock at altitudes of about 1000m. The mean maximum temperature of the hottest month is about 28°C, and the mean minimum of the coldest month about 5°C. A few frosts occur and, very rarely, there are short falls of snow. The mean annual rainfall is about 1200mm with a pronounced summer peak. The species generally occurs on its own as scattered trees amongst other eucalypts such as *E. punctata* and *E. andrewsii*.

Conservation

The main stands occur in Girrawheen National Park in Queensland and, together with the nature of the sites on which the stands occur, the species is at little risk.

Special Features

The distinctness of the species and its highly restricted distribution, is in accord with the relict nature of this and a number of other species referred to elsewhere. It is planted quite extensively as an ornamental.

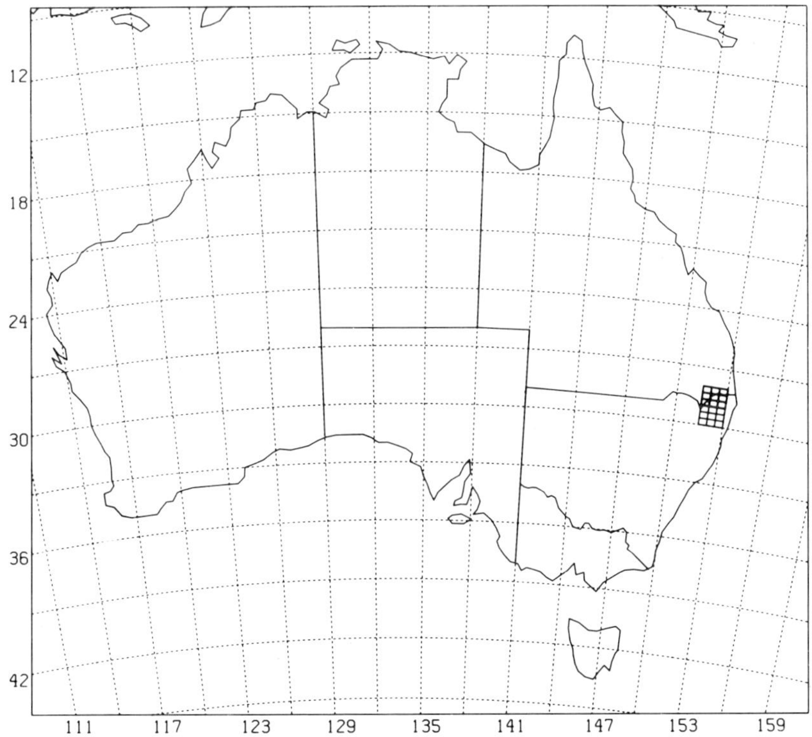
12
18
24
30
36
42
111
117
123
129
135
141
147
153
159

Eucalyptus sepulcralis F. Muell. **Blue Weeping Gum** (MADAC F.T.S.—)

A small drooping tree found south of Ravensthorpe in Western Australia.

Distribution

The recorded localities of this species are the Ayers Range northwest of Hopetoun and adjoining localities in the Fitzgerald River National Park.

Description

It is a small slender tree up to about 8m in height with drooping glaucous branches and bright green leaves. The bark is smooth and white. The flower clusters are in sevens and the fruit is rather large being more than 3cm by 2cm. The fruit is contracted at the orifice and somewhat costate when dry.

Ecology

Blue Weeping gum occurs in ridgetops in rather undulating moderately hilly country on light sandy soils derived from granite. The mean maximum temperature of the warmest month is about 23°C and the mean minimum temperature of the coldest month about 5°C. It stands out above a wide array of sclerophyll shrubs such as *Banksia, Grevillea, Hakea and Acacia* which are characteristic of the shrub heath of the area.

Conservation

Almost all the main stands are within the Fitzgerald River National park where preservation of the vegetation is one of the aims of management. In spite of its very limited distribution therefore, it is well preserved at present.

Special Features

This species is one of the small group of eucalypts with very pendulous branches and open sketchy foilage. It is one of the small group of species within the Monocalyptus sub-genus endemic to Western Australia.

Photo—National Botanic Gardens—M. D. Crisp

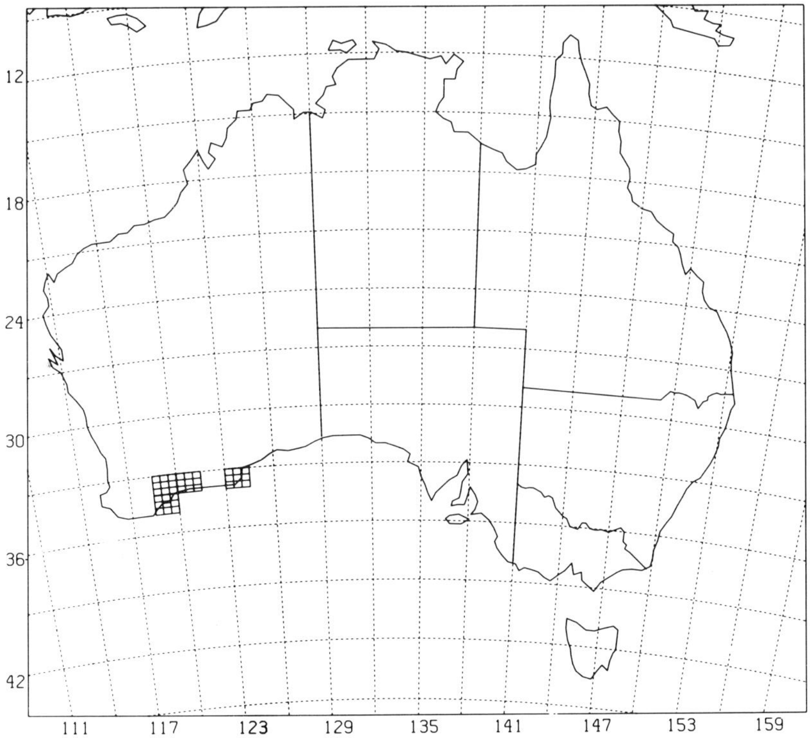

Eucalyptus squamosa Deane et Maiden **Scaly Bark** (SIQ:A F.T.S. 35)

E. squamosa is a small rough-barked tree of the central coast of New South Wales.

Distribution

This species occurs in small stands on sandstone sites in the vicinity of Sydney, particularly Ku-ring-gai, Blue Mountains and Royal National parks.

Description

This small rather straggly tree is found on hard sites and typically it has a rather crooked or leaning stem. It grows to about 15m although the best trees may reach 20m in height. The bark is rough and scaly and occurs in rectangular patches over the main trunk and larger branches. The buds are pointed and fruits small with fine distinctly exserted valves. The adult leaves are rather dull in colour. This species is related to some mallee species of the *Subulatae* group which are mostly found in Western Australia.

Ecology

Scaly Bark occurs on broad ridgetops and lower plateaux which have underlying Hawkesbury sandstone. It is found in latitude 34°S, mostly close to Sydney. The mean maximum temperature of the hottest month is about 30°C and the mean minimum of the coldest month about 7°C. There are very few frosts and the mean annual rainfall is around 1100mm, although it is less at the western limit of the species at the foot of the Blue Mountains. There is a summer peak in rainfall pattern. The soils are mainly shallow, sandy loams and at times are skeletal. It is associated with other eucalypts such as *E. haemastoma, E. racemosa, E. gummifera, E. eximia* and *E. sieberi* with many sclerophyll shrubs.

Conservation

The most extensive stands, which in any event are not very large, occur in national parks and its preservation seems assured.

Special Features

E. squamosa is of special interest since, along with *E. bakeri* and *E. pachycalyx* of north Queensland, it is one of a considerable number of species with bisected cotylendons which are found mainly in southern and western Western Australia. The existence of the small stands of this species suggests that is a detached eastern relict occurrence from former but very ancient and widespread distribution of forerunners of the group *Subulatae*. It is a reverse situation from the distribution pattern of *E. remota*.

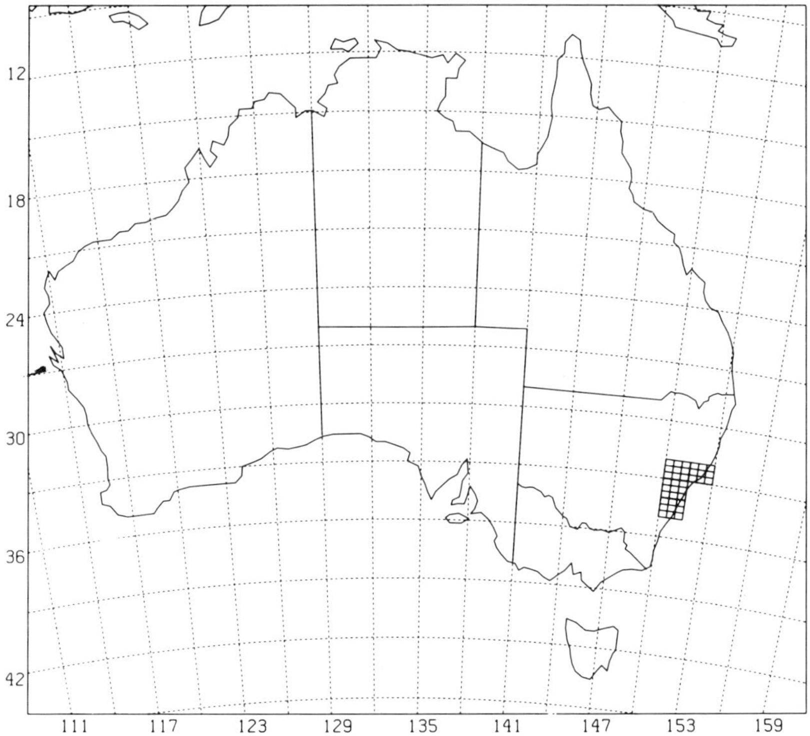
12
18
24
30
36
42
111
117
123
129
135
141
147
153
159

Eucalyptus steedmanii Gardner **Steedmans Gum** (SIDCF F.T.S.—)

A small upright tree found in the Norseman area of Western Australia.

Distribution

E. steedmanii occurs in two or three small stands some distance east of Hyden towards Norseman in Western Australia.

Description

Steedmans Gum is a slender-stemmed tree, occasionally mallee, reaching 8-9m in height. It has erect branches and a small compact crown. The adult leaves are rather obtuse and thick with a lustrous appearance. The fruit is distinctive being markedly four-angled and winged. This species resembles *E. forrestiana* though the former is smaller and the two species are not closely related.

Ecology

E. steedmanii occurs in small rather dense stands in a limited area. It is found in about latitude 33°S and at an altitude of 300m. The mean maximum temperature of the hottest month is about 28°C and the mean minimum temperature of the coldest month is about 3°C. There are occasional frosts and the mean annual rainfall is about 400mm with a marked winter peak. It occurs in two or three small stands in which it is the predominant species but a number of other eucalypts are in the same vicinity. It is associated with sclerophyllous undergrowth shrubs. The soil is light and sandy texture with a considerable amount of lateritic gravel and some massive laterite.

Conservation

This species occurs in areas subject to mineral exploration and if this led to mining there would be a threat to the survival of the species. The two or three stands which are each of less than 3 hectares could be eliminated entirely. Reservation of a representative stand is very desirable. Developed access in this part of Western Australia is limited and it is quite possible that intensive survey could reveal more stands than are known at present.

Special Features

E. steedmanii had not been seen in the field since 1938 and it was considered by some possibly to be extinct. However, its rediscovery recently shows that the stands are in good condition and that regeneration is occuring actively. Regeneration is particularly active where there has been some disturbance, for example, due to transect mining survey lines.

12
18
24
30
36
42
111
117
123
129
135
141
147
153
159

Eucalyptus stenostoma Johnson & Blaxell **Jilliga Ash** (MAKIA F.T.S. 189)

This species is found on a number of sites in the dissected Tuross River area of southern New South Wales.

Distribution

It is found on ridges and spurs in the general vicinity of Nerrigundah, spurs of Wadbilliga Mountain and a number of other sites in this general locality.

Description

Jilliga Ash is a fairly tall tree, up to 25m in height with a rough bark over much of its bole and often with a leaning curved trunk. The branchlets, buds and fruits are highly glaucous which is a distinctive feature of the species. It is superficially like *E. fraxinoides* with which it has at times been confused. The fruit is globular with the orifice very much constricted and giving a small opening to the sunken valves.

Ecology

Jilliga Ash occurs in about latitude 36°S and at altitudes of about 650m. The mean maximum temperature of the hottest month is about 26°C and the mean minimum temperature of the coldest month is about 0°C. The mean annual rainfall is around 850mm with a good distribution tending towards a summer peak. It occurs on rocky areas but in a general zone of eucalypt sclerophyll forest where the soils are poor and shallow derived from sedimentary rocks such as slate or shale. It occurs in almost pure stands with surrounding species commonly being *E. sieberi, E. globoidea, E. muellerana* and *E. cypellocarpa*.

Conservation

It occurs mostly in state forest which is managed for continuous production from eucalypts. It is desirable that the management prescription of these state forests should include provision for maintaining the species. This would not be difficult under the ordinary system of management as this species regenerates quite adequately.

Special Features

Jilliga Ash was collected first in 1928 by F.M. Bailey but it was misidentified as a variant of *E. fraxinoides*. Subsequently re-examination showed that although related to *E. fraxinoides* it is clearly a distinct species. It differs in a number of features including colour of seed. Jilliga Ash is a quite restricted species although the existing stands probably amount to more than a hundred hectares which is distributed in small patches. The area in which the species occurs is very rough and dissected and difficult to explore. It is likely there are more stands existing than have so far been recorded.

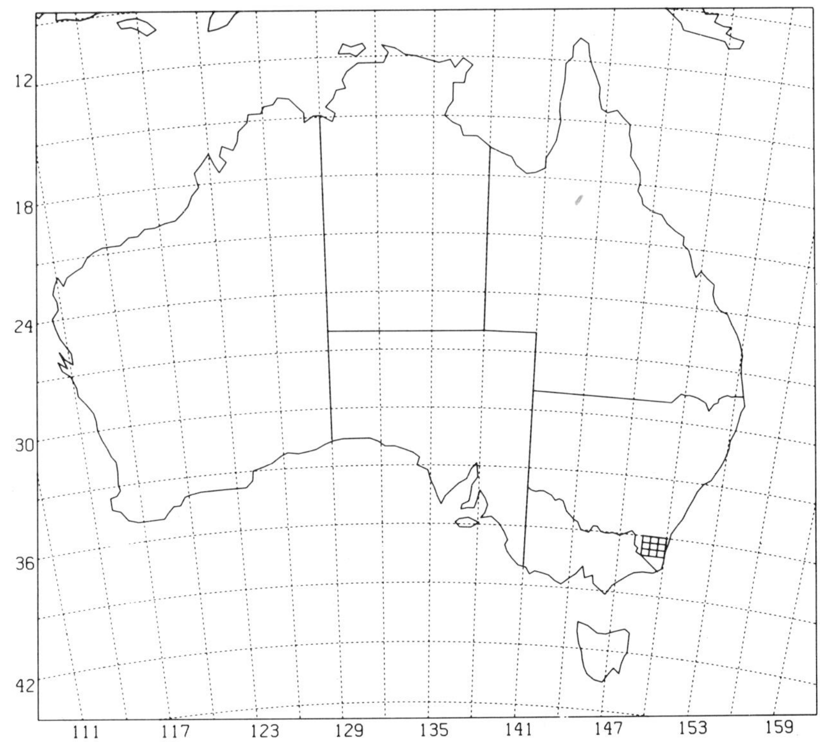
12
18
24
30
36
42
111
117
123
129
135
141
147
153
159

Eucalyptus stoatei Gardner **Stoates Moort** (SLOBA F.T.S.—)

This is a small tree found in the general vicinity of Ravensthorpe, Western Australia.

Distribution

It is found in a number of localities particularly on the Jerdicuttup Creek in the vicinity of Kundip near Badalup in Western Australia. This site is between Ravensthorpe and Esperance.

Description

It is a small erect tree up to 10m in height with a smooth, grey-green thin bark and with dense branches with leaves clustered towards the end. The leaves are mostly obtuse, quite thick, coriaceous, dark green and shining. The flowers occur singly in the axils of leaves and the fruits are quite large and very strongly ribbed. The buds are generally scarlet and the fruit deep yellow when it is ripening.

Ecology

The species occurs on undulating sites with lateritic gravel and some sandy soil. it occurs in about latitude 33°S at an altitude of about 300m. The mean maximum temperature of the hottest month is about 25°C and the mean minimum temperature of the coldest month is about 5°C. There are occasional frosts and a winter rainfall of about 350mm per annum. It occurs in small rather dense stands with some associated shrubs. There are a number of other eucalypt species in the same general vicinity.

Conservation

It is likely that some stands of *E. stoatei* have been cleared from land for grazing. The existing stands are limited and not under much threat from present land use patters but could be eliminated by mining or similar activity. Additional survey is necessary to establish whether other stands exist. It is desirable that reserves should be declared to include some of the best stands.

Special Features

E. stoatei is a quite restricted endemic species of Western Australia and is of considerable ornamental value because of the colour of buds and fruits. It is one of the few eucalypts in which the flowers occur singly on a peduncle in the axil of the leaf.

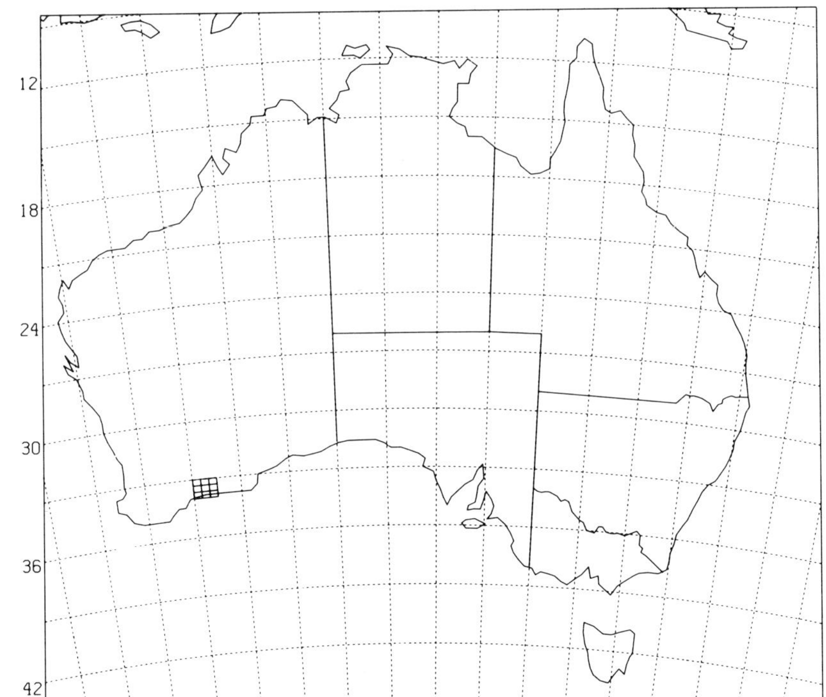
12
18
24
30
36
42
111
117
123
129
135
141
147
153
159

Eucalyptus sturgissiana Johnson et Blaxell **Ettrema Mallee** (SPIAF F.T.S. 115)

This is a small mallee of the central south coast of New South Wales.

Distribution

This species occurs in a very limited area about 20 to 50km southwest of Nowra in NSW at about latitude 35°S. It occurs only in a few small patches.

Description

Ettrema Mallee is a small mallee with a number of stems which grow up to about 5m. It has a straggly appearance. The leaves are rather thick and the stems light-coloured. The juvenile leaves are opposite, ovate and become connate and perfoliate, a feature which is shared by few other species. The buds and fruits are in 7's with the fruit small.

Ecology

This species occurs at altitudes from about 250 to 700m in an area of dissected sandstone plateaux. The mean maximum temperature of the higher sites in the hottest month is about 28°C and the mean minimum of the coolest month about 2°C. It is warmer at the lower altitudes. There are probably around 20 frosts a year. The mean annual rainfall is about 800mm and is well distributed throughout the year. It occurs in small pure stands with very few individual plants in each of the separate populations. Ground cover in the stands is of grasses and sedges which will tolerate seasonally wet conditions. On adjacent sites there are stunted *E. gummifera* and also *E. punctata*. The sites of occurrence are flat zones associated with massive outcrops of horizontally bedded Permian sandstone. The soils are associated with impeded drainage and become very dry during rainless periods.

Conservation

Most of the known stands are within the Morton National Park and are therefore assured of preservation.

Special Features

Ettrema Mallee is of interest because it is a distinct species of very limited occurrence and therefore may be a relict occurrence.

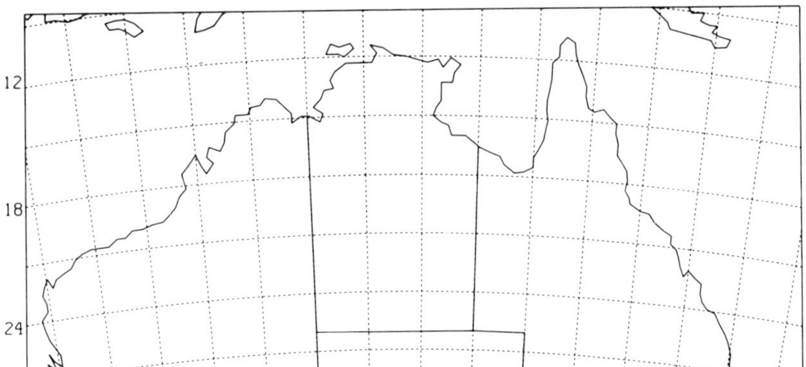

12
18
24
30
36
42
111
117
123
129
135
141
147
153
159

Eucalyptus tetrapleura Johnson **Square-fruited Ironbark** (SUV:C F.T.S. 170)

This is a moderate sized tree of the central north coast of New South Wales.

Distribution

This species occurs on coastal lowlands and adjacent foothills in the vicinity of Grafton and in the southern parts of the Casino district. Its distribution is within an area of about 100km long and 40km wide and it reaches to within about 20km of the sea. Its distribution is patchy within this area which is generally in latitude about 29.25°S.

Description

It is a moderate sized tree reaching 30m in height. It has a very thick furrowed ironbark and even on poor sites the trunk remains straight. The flower clusters are in 7's. The most distinctive feature is the somewhat winged fruit which is quadrangular in cross-section.

Ecology

Square-fruited Ironbark mainly occurs at altitudes from 10 to 150m, but occasionally may reach over 300m. The mean maximum temperature of the hottest month is about 31°C and the mean minimum temperature of the coolest month about 4°C. Frosts are rare. Mean annual rainfall is around 1100mm with a moderate summer maximum. In the lowlands it occurs mainly on low rises on minor drainage lines. The soils have poor fertility and have a heavy clay subsoil and poor internal drainage. The soil surface layers may be lateritic or sandy whereas further inland the soils have better structure. Generally the parent rocks are sandstones or shales. Other eucalypts found in association with this ironbark, or close by, include *E. maculata, E. gummifera* and *E. bancroftii* and associated shrubs include *Daviesia* and *Melaleuca*. Also found in the same general vicinity are *E. fibrosa, E. racemosa* and *E. signata* as well as a number of species of *Angophora*.

Conservation

Many of the known occurrences are on private property but some are in Newfoundland State Forest. Recent land development has led to clearing of some of the stands. Effective protection of the species will depend on documenting the distribution of the species, especially in reserves and ensuring that where it occurs in state forest, management prescriptions allow for its persistence.

Special Features

This ironbark is unusual amongst the ironbarks because of the shape of the fruit. The south coast of NSW form of *E. paniculata* shows some tendency to a quadrangular shape but it is never as strongly costate as in *E. tetrapleura*.

12
18
24
30
36
42
111
117
123
129
135
141
147
153
159

Eucaluyptus triflora (Maiden) Blakely **Pigeon House Ash** (MAKIC F.T.S. 113)

This is a small tree found in the central south coast of New South Wales.

Distribution

It is found in a very few localities some 20-40km west of Ulladulla on dissected sandstone country. The stands are generally in the region of Pigeon House Mountain and the Endrick River.

Description

Generally it is a small slender tree up to about 12m in height, although occasionally it is somewhat taller. In some sites it may be reduced to a mallee form. The fruits are somewhat urceolate and the buds small and squat. It has shiny leaves and the bark is mostly smooth. The most distinctive feature is that the buds occur in 3's which is an unusual feature for a species of Monocalyptus sub-genus.

Ecology

Pigeon House Ash occurs in about latitude 35°S and at altitudes of from 700-800m. The mean maximum temperature of the warmest month is about 28°C and the mean minimum temperature of the coolest month about 3°C. It is likely there are about 15-20 frosts a year. The mean annual rainfall is about 1100mm with a moderate summer peak. it grows on elevated plateau-like areas along broad ridgetops which is often where there are massive outcrops of Permian sandstones. The tree grows in fissures between the boulders on rather poor well drained sandy soils. This species occupies sites which apparently are too poor to support the adjacent species which are commonly *E. sieberi, E. agglomerata* and *E. smithii.*

Conservation

The known occurrences of Pigeon House Ash are mainly within the Moreton National Park and therefore preservation should be assured.

Special Features

This species is a very restricted endemic of NSW which is interesting because of its unusual morphology within the Monocalyptus sub-genus. it is related to *E. fraxinoides* and *E. dendromorpha*. However its specific status appears to be fully justified.

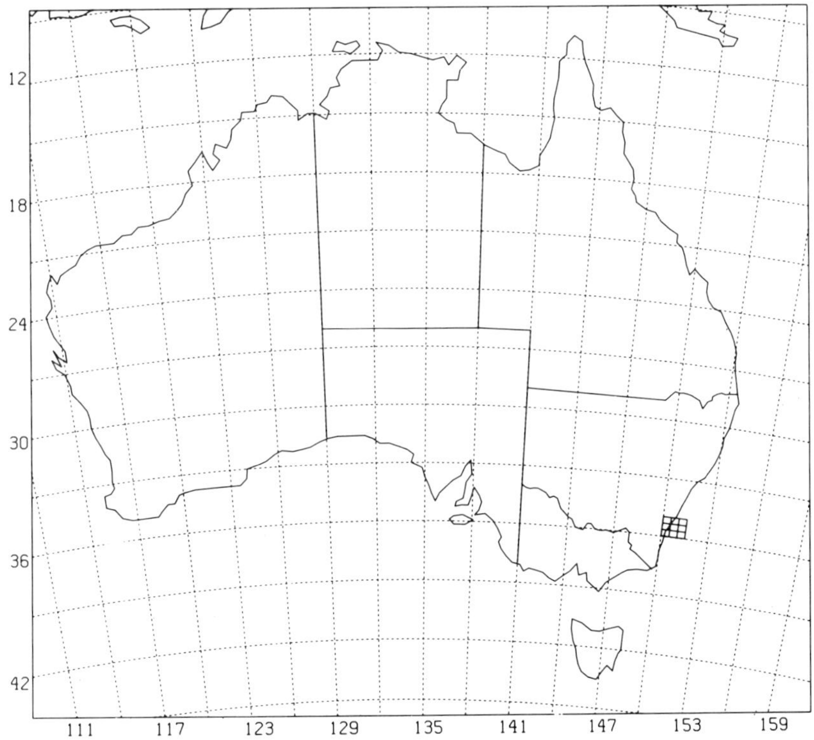
12
18
24
30
36
42
111
117
123
129
135
141
147
153
159

Eucalyptus woodwardii Maiden **Woodwards Blackbutt** (SLE:A F.T.S.—)

This is a small straight tree east of Kalgoorlie in Western Australia.

Distribution

This species occurs in small stands at 3 or 4 localities east of Kalgoorlie between Karonie and Xanthus. Access in this area of Western Australia is poorly developed and from the way in which the known stands occur it is quite likely that other stands exist. Nevertheless, it is a species of quite restricted occurrence although the various known occurrences are distributed over several hundred kilometres.

Description

E. woodwardii is a straight glaucous-leaved tree about 17m tall. It has smooth bark except for a little persistent rough bark near the butt. The branches are often very pendulous and at times sweep the ground. The adult leaves are very thick, coriaceous and rather large. The flower clusters are 7-flowered, lemon-yellow to orange-yellow in colour. The fruits are very glaucous, fairly large and campanulate.

Ecology

Woodwards Blackbutt occurs in pure stands on low sandy ridges. it is difficult to discern the site differences which account for this species distribution pattern in an area where a number of other species are common. The mean maximum temperature of the hottest month is about 30°C and the mean minimum temperature of the coldest month about 2°C. Several frosts occur each year and the rainfall is about 250mm or a little less with a pronounced winter peak.

Conservation

There is not much risk of clearing for land development because of the adverse climate of the area. It is desirable that some stands should be specifically protected in case of some unforseen land development which could lead to the species elimination.

Special Features

Woodwards Blackbutt is a species of considerable ornamental value because of the rather large bright yellow flowers and it has been planted quite extensively in southern Australia. It has hybridized under cultivation with *E. torquata*, Coral Gum, and the resultant hybrid known as 'Torwood' is a popular ornamental tree.

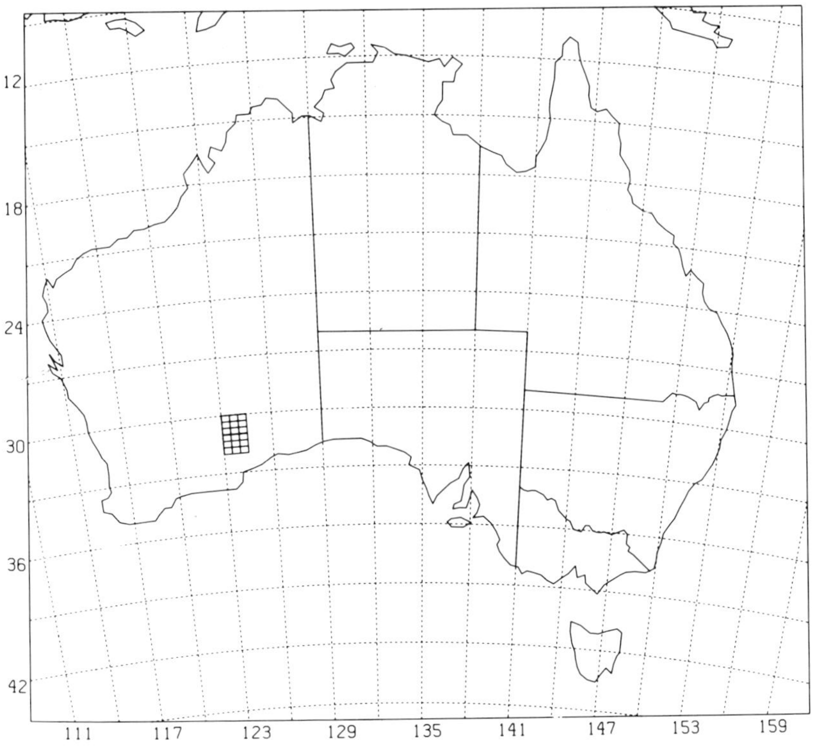
12
18
24
30
36
42
111
117
123
129
135
141
147
153
159

Eucalyptus yarraensis Maiden et Cambage **Yarra Gum** (SPEAC F.T.S. 76)

Yarra Gum is a medium sized umbrageous tree from southern Victoria in the general vicinity of Melbourne.

Distribution

The original distribution of *E. yarraensis* is difficult to determine as it is found on fertile sites and most of it has been cleared for agriculture. There are small surviving stands near Healesville, Lilydale and also in the vicinity of Ballarat.

Description

This species is a round-headed moderate sized tree up to about 18m in height with a spreading crown. The buds and fruits are in clusters of 7. The leaves have rather long petiole and are often somewhat ovate in the intermediate stage. The adult leaves are rather broadly ovate. The bark is rough, flaky and fibrous throughout the trunk and the main branches. The bark is the main distinguishing feature from the two most closely related species, *E. camphora* and *E. ovata*. Yarra Gum also has some affinity with *E. aggregata* but the leaves of that species are much narrower and the juvenile leaves different.

Ecology

E. yarraensis occurs in areas of reasonably high rainfall on good soils near river banks and along valley bottoms. it is found over an east-west distance of 400km at a latitude of about 38°S and an altitude of about 100m. The mean maximum temperature of the warmest month is about 26°C and the mean minimum temperature of about 5°C. There are up to about 15 frosts a year and the mean annual rainfall is variable from about 700-1000mm with a weak winter maximum. On alluvial soils it forms a woodland with grassy undergrowth and is often associated with *E. viminalis* and *E. obliqua*.

Conservation

Because of the topography and quality of the soils on which this species occurs, most stands of *E. yarraensis* have been cleared. The species was probably never numerous and there are now mere remnants of the species in scattered localities throughout its original range. Because most of the sites on which it occurs are on privately owned farmland the surviving stands are only to be found in corners of paddocks and along roadside verges. It is therefore in a highly precarious position and one of the most threatened of the eucalypt species. Current detailed survey being carried out by interested individuals should lead to formulation of a management plan to ensure the species survival.

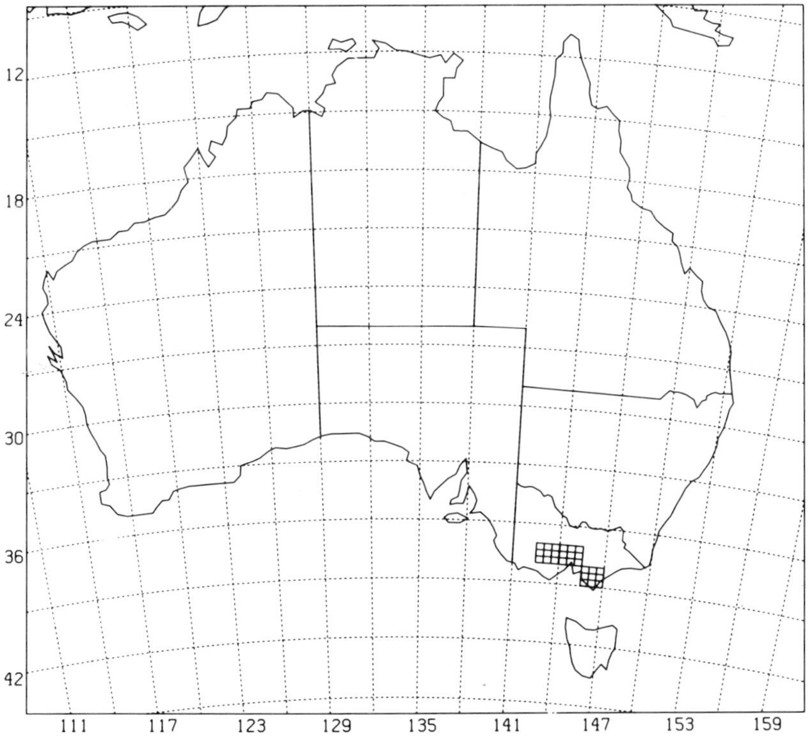
12
18
24
30
36
42
111
117
123
129
135
141
147
153
159

GLOSSARY

campanulate: bell shaped.
coriaceous: thick, leathery.
costate: ribbed.
cruciform: cross shaped.
decorticate: to shed by separation of outer layers of bark.
dilated: expanded and flattened.
disc: top of fruit surrounding valves.
discolorous: distinctly different colour of upper and lower leaf surfaces.
flexuose: springy.
glaucous: bluish white colour or bloom due to surface covering of fine wax particles.
globoid: round or spherical.
keel: prominent ridge.
lanceolate: the shape of the blade of a lance.
ovate: the shape of the outline of an egg.
paniculate: a flower cluster with two orders of racemose branching.
peduncle: the common stalk to which a number of separate flowers are attached.
pendulous: hanging.
perfoliate: opposite leaves joined at the base so that the stem is encircled completely by leaf tissue.
rostrate: beak shaped.
sessile: without a stalk or petiole.
stellate: star shaped.
tessellated: like a tiled surface.
truncate: abruptly cut off.
umbrageous: spreading or shady.
urceolate: urn shaped.
verrucose: warty.

COMMON NAME INDEX OF ENDANGERED SPECIES LISTED IN THE TEXT

RodenPrint